쉽게 배우는 해양실무영어

Offshore English

감수 박진희 저자 이정은

도서출판 GS인터비전

Introduction

Due to the continuing downturn in shipping industries, the shipbuilding sector has been badly affecting these days as financially squeezed European ship owners are still hesitant to place orders or cancel orders.

However, since Korea's large shipbuilders have capability in producing offshore plants, they will be less affected by the global economic slowdown in spite of the declining demand for merchant ships.

Having considered that offshore is saving Korean shipbuilders, this book has been written to provide the basic knowledge on offshore industry and communication skill.

So, the content of this book is mainly focused on building basic knowledge on the offshore industry.

Over the writing period, I have been given lots of advice from my co-workers and seniors who contribute to the shipbuilding industries and I thank them all. And, I thank all my students who willingly gave me their own pictures. I am also grateful to my co-worker, Jin Hee Park who helped the English grammar in this book. Also, my sincere thanks go to Jung ji, Young, President of Koje University, who encourage and give me the chance to write this book. Finally, I always thank to my family and God to support me all the time.

2012 / 09 / 10
JeongEun-Lee

Section	Subject	Page
	Unit 1	1
Lecture A	How Has Offshore Drilling Been Developing?	2
Lecture B	After you read	4
Lecture C	Language Focus	5
Lecture D	Useful expression- quick/quickly	6
Lecture E	Speak out	7
Lecture F	Episode on offshore!!	8
Outro	Whale Oil Versus The Others	9
Pops English	The Rose	10
	Unit 2	13
Lecture A	How Do FPSOs Work?	14
Lecture B	After you read	16
Lecture C	Language Focus	17
Lecture D	Useful expression- one~ the other	18
Lecture E	Speak out	19
Lecture F	Episode on offshore!!	20
Outro	Pipeline transport	21
Pops English	To sir with love	22

Section	Subject	Page
	Unit 3	25
Lecture A	How Do Semi-submersible Drilling Rigs Work?	26
Lecture B	After you read	28
Lecture C	Language Focus	29
Lecture D	Useful expression- as soon as	30
Lecture E	Speak out	31
Lecture F	Episode on offshore!!	32
Outro	What is barrel?	33
Pops English	My heart will go on	34
	Unit 4	37
Lecture A	What is a Jack Up Drilling Rig?	38
Lecture B	After you read	40
Lecture C	Language Focus	41
Lecture D	Useful expression- the deepest thing	42
Lecture E	Speak out	43
Lecture F	Episode on offshore!!	44
Outro	Burning ice	45
Pops English	In dreams	46

Section	Subject	Page
	Unit 5	49
Lecture A	What is a TLP?	50
Lecture B	After you read	52
Lecture C	Language Focus	53
Lecture D	Useful expression- every/each	54
Lecture E	Speak out	55
Lecture F	Episode on offshore!!	56
Outro	How the seas were named (1)	57
Pops English	Just Called To Say I Love You	58
	Unit 6	61
Lecture A	What Is a Tension Spar Platform?	62
Lecture B	After you read	64
Lecture C	Language Focus	65
Lecture D	Useful expression- both/either/neither	66
Lecture E	Speak out	67
Lecture F	Episode on offshore!!	68
Outro	How the seas were named (2)	69
Pops English	Puff, the magic dragon	70

Section	Subject	Page
	Unit 7	73
Lecture A	What Is a Drillship?	74
Lecture B	After you read	76
Lecture C	Language Focus	77
Lecture D	Useful expression- so/such	78
Lecture E	Speak out	79
Lecture F	Episode on offshore!!	80
Outro	The story of the Plimsoll Mark	81
Pops English	Just the way you are	82
	Unit 8	85
Lecture A	What Is a Shuttle Tanker?	86
Lecture B	After you read	88
Lecture C	Language Focus	89
Lecture D	Useful expression- It is said that~	90
Lecture E	Speak out	91
Lecture F	Episode on offshore!!	92
Outro	Viking tales, true or false	93
Pops English	Rhythm of falling rain	94

Section	Subject	Page
	Unit 9	97
Lecture A	What Is a FLNG?	98
Lecture B	After you read	100
Lecture C	Language Focus	101
Lecture D	Useful expression- speak to you/ mention it	102
Lecture E	Speak out	103
Lecture F	Episode on offshore!!	104
Outro	Viking laws	105
Pops English	Sorry seems to be the hardest word	106
	Unit 10	109
Lecture A	How Does Offshore Pipeline Installation Work?	110
Lecture B	After you read	112
Lecture C	Language Focus	113
Lecture D	Useful expression - ~ing/P.P phrases	114
Lecture E	Speak out	115
Lecture F	Episode on offshore!!	116
Outro	Seven Sisters	117
Pops English	Don't forget to remember	118

Section	Subject	Page
	Unit 11	121
Lecture A	How Do Mooring Systems Work?	122
Lecture B	After you read	124
Lecture C	Language Focus	125
Lecture D	Useful expression –expressions + ~ing	126
Lecture E	Speak out	127
Lecture F	Episode on offshore!!	128
Outro	The organization of the petroleum exporting countries	129
Pops English	You raise me up	130

Section	Subject	Page
	Unit 12	133
Lecture A	What Is Dynamic Positioning?	134
Lecture B	After you read	136
Lecture C	Language Focus	137
Lecture D	Useful expression – If I were you	138
Lecture E	Speak out	139
Lecture F	Episode on offshore!!	140
Outro	Marine Mineral	141
Pops English	Love of my life	142

Section	Subject	Page
	Unit 13	145
Lecture A	What Is the Turret Mooring System?	146
Lecture B	After you read	148
Lecture C	Language Focus	149
Lecture D	Useful expression – how to do	150
Lecture E	Speak out	151
Lecture F	Episode on offshore!!	152
Outro	Tidal Energy &Current Energy	153
Pops English	I have a dream	154
	Unit 14	157
Lecture A	What Is a LNG Re-gasification Vessel?	158
Lecture B	After you read	160
Lecture C	Language Focus	161
Lecture D	Useful expression - prepositions	162
Lecture E	Speak out	163
Lecture F	Episode on offshore!!	164
Outro	Six day war and advent of VLCC	165
Pops English	I will follow him	166

Section	Subject	Page
	Unit 15	169
Lecture A	How do ROVs work?	170
Lecture B	After you read	172
Lecture C	Language Focus	173
Lecture D	Useful expression – Phrasal Verbs Ⅰ	174
Lecture E	Speak out	175
Lecture F	Episode on offshore!!	176
Outro	Offshore wind power	177
Pops English	You are my sunshine	178
	Unit 16	181
Lecture A	How does blowout control work?	182
Lecture B	After you read	184
Lecture C	Language Focus	185
Lecture D	Useful expression - Phrasal Verbs Ⅱ	186
Lecture E	Speak out	187
Lecture F	Episode on offshore!!	188
Outro	Deepwater horizon oil spill	189
Pops English	Let it be	190

UNIT 1

How
Has Offshore Drilling Been Developing?

어떻게 해양 개발이 시작됐는가?

Lecture A How Has Offshore Drilling Been Developing?

Around 1891, the first submerged oil wells were drilled from platforms built on piles in the fresh waters of the Grand Lake St. Marys in Ohio. The wells were developed by small local companies such as Bryson, Riley Oil, German-American, and Banker's Oil. Around 1896, the first submerged oil wells in salt water were drilled in the portion of the Summerland field extending under the Santa Barbara Channel in California. The wells were drilled from piers extending from land out into the channel. Other notable early submerged drilling activities occurred on the Canadian side of Lake Erie in the 1900s and Caddo Lake in Louisiana in the 1910s. Shortly thereafter wells were drilled in tidal zones along the Texas and Louisiana gulf coast. The Goose Creek Oil Field near Baytown, Texas is one such example. In the 1920s drilling activities occurred from concrete platforms in Venezuela's Lake Maracaibo.

One of the oldest subsea wells is the Bibi Eibat well, which came on stream in 1923 in Azerbaijan. The well was located on an artificial island in a shallow portion of the Caspian Sea. In the early 1930s, the Texas Co., later Texaco (now Chevron) developed the first mobile steel barges for drilling in the brackish coastal areas of the Gulf of Mexico. In 1937, Pure Oil (now Chevron) and its partner Superior Oil (now ExxonMobil) used a fixed platform to develop a field 1 mile offshore of Calcasieu Parish, Louisiana in 14 feet of water.

In 1946, Magnolia Petroleum (now ExxonMobil) drilled at a site 18 miles off the coast, erecting a platform in 18 feet of water off St. Mary Parish, Louisiana. In early 1947, Superior Oil erected a drilling and production platform in 20 feet of water some 18 miles off Vermilion Parish, La. But it was Kerr-McGee Oil Industries (now Anadarko Petroleum), as operator for partners Phillips Petroleum (ConocoPhillips) and Stanolind Oil & Gas (BP) that completed its historic Ship Shoal Block 32 well in October 1947, months before Superior actually drilled a discovery from their Vermilion platform farther offshore. In any case, that made Kerr-McGee's well the first oil discovery drilled out of sight of land.

When offshore drilling moved into deeper waters of up to 100 feet, fixed platform rigs were built, until demands for drilling equipment was needed in the 100- to 400-foot depth of the Gulf of Mexico, the first jack-up rigs began appearing from specialized offshore drilling contractors such as forerunners of ENSCO International.

oil well 유정 a well that yields or has yielded oil **drilling platform** 작업용 플랫폼 a structure, either fixed to the sea bed or mobile, which supports the machinery and equipment (the drilling rig), together with the stores, required for digging an offshore **artificial** 인공적인 contrived by art rather than nature **jack-up rig** 잭업 리그 a type of mobile platform that consists of a buoyant hull fitted with a number of movable legs, capable of raising its hull over the surface of sea

The first semi-submersible resulted from an unexpected observation in 1961. Blue Water Drilling Company owned and operated the four-column submersible Blue Water Rig No.1 in the Gulf of Mexico for Shell Oil Company. As the pontoons were not sufficiently buoyant to support the weight of the rig and its consumables, it was towed at a draught mid-way between the top of the pontoons and the underside of the deck. It was noticed that the motions at this draught were very small, and Blue Water Drilling and Shell jointly decided to try operating the rig in the floating mode. The first purpose-built drilling semi-submersible Ocean Driller was launched in 1963. Since then, many semi-submersibles have been purpose-designed for the drilling industry mobile offshore fleet.

The first offshore drillship was the CUSS 1 developed for the Mohole project to drill into the Earth's crust. As of June, 2010, there were over 620 mobile offshore drilling rigs (Jackups, semisubs, drillships, barges) available for service in the competitive rig fleet.

The world's deepest platform is currently the Perdido in the Gulf of Mexico, floating in 2,438 meters of water. It is operated by Royal Dutch Shell and was built at a cost of $3 billion.

semi-submersible 반잠수식구조물 a specialized marine vessel with good stability and seakeeping characteristics **draught** 흘수 the depth of a vessel's keel below the surface **observation** facts learned by observing **pontoon** 폰툰 a floating structure that serves as a dock or to support a bridge **sufficiently** 충분하게 to a sufficient degree **buoyant** 부력있는 tending to float on a liquid or rise in air or gas

Lecture B **After You Read**

1. Fill in the blanks based on this article.

   (a) In the early 1930s, Chevron developed________________ for drilling.

   (b) When offshore drilling moved into __________ of up to 100 feet, ____________were built, until demands for drilling equipment was needed in the ______________of the Gulf of Mexico.

2. What is the topic of this article?

   (a) Issues on the offshore marketing

   (b) History of Drilling development in offshore

   (c) Improvement of drilling rigs

   (d) Pros and cons of offshore business

3. When were the first submerged oil wells drilled in sea water?

   (a) Around 1943

   (b) Around 1896

   (c) Around 1891

   (d) Around 1937

4. Which one of the following is **not** true?

   (a) The first semi-submersible resulted from an unexpected observation in 1961.

   (b) The pontoons were sufficiently buoyant to support the weight of the rig and its consumables.

   (c) The first offshore drillship was the CUSS 1 developed for the Mohole project to drill into the Earth's crust.

   (d) One of the oldest subsea wells is the Bibi Eibat well, which came on stream in 1923 in Azerbaijan.

Lecture C	Language Focus

1. Match the words with the definitions, then complete the definition with the given words.

Drill Platform	●	a structure, either ________ to the sea bed or mobile, which supports.
Semi-submersible	●	a type of ____________ platform that consists of a ____________ fitted with a number of movable legs
Jack up rigs	●	a specialized vessel with good ______________ and seakeeping characteristics

fixed   stability   mobile   buoyant hull

2. Complete the sentences using the following words.

resulted   out of sight   towed

1) The first semi-submersible _____________ from an unexpected observation in 1961.
2) It was _______ between locations at a draught mid-way between the top of the pontoons and the underside of the deck.
3) In any case, that made Kerr-McGee' s well the first oil discovery drilled _______________ of land.

3. Complete the sentences using the following words.

1) Other notable early submerged drilling activities ________ on the Canadian side of Lake Erie.
   a) occurred   b) occurring   c) had occurred   d) occur
2) ___________ then, many semi-submersibles have been purpose-designed for the drilling industry mobile offshore fleet.
   a) In addition   b) Since   c) Why   d) Because

Lecture D Useful Expressions - quick/quickly

A. quick/quickly

She speaks **perfect Korean**.	She **speaks** Korean **perfectly**.
adjective + noun	verb + noun + adverb

Many adverbs are formed from an adjective + **-ly**

Adjective : quick	moral	simple	heavy	happy	full	true
Adverb : quickly	morally	simply	heavily	happily	fully	truly

Not all words that end is **–ly** are adverbs . Some adjectives end with **–ly**, too. For example:

costly elderly friendly likely ugly lonely lovely silly lively

B. Hard/hardly : be careful of these meanings

hard	I work **hard** everyday.
hardly	I can **hardly** believe what you are saying.
most	That is the **most** touching story.
mostly	My friends are **mostly** friendly.
late	I got up **late** yesterday. / I came home **late** last night.
lately	I haven't had enough sleep **lately**. /Have you met him **lately**?
near	The Christmas is coming **near**. /Come **nearer**.
nearly	The sun is **nearly** down.
high	Raise your glasses **high**./ Fly **high** to the blue sky like an eagle.
highly	He is a **highly** educated man. / She is a **highly** placed office.
close	He came to me and sat **close**. / The dog came **close** to her.
closely	See these signposts **closely**. /Look at your face **closely**.
great	Well done. You did **great**. / The project is going **great**.
greatly	He is **greatly** respected by all. / You will be **greatly** missed.

Lecture E **Speak Out!!**

1. Talk with your partner. What would you say in these situation? Complete the sentences using the words below.

~~nervous(ly)~~	true(truly)	love(ly)	perfect(ly)

	Situation	You say
	**Your friend is asking you about your exam.**	**Do you usually feel <u>nervous</u> before exams?**
	You are walking in the park with your friend.	Look at that!!! This is so ____________ !!
	You are watching a ballet show on TV. He is very good.	He is a ___________ dancer. He dances ______________ .
	You are playing football, and you've finally won.	I am __________ happy! It is a ____ winning!

2. Have a conversation with your partner using the words below.

late(ly)	true(truly)	hard(ly)	early

	Hey, Hanna. You are ____ for my class again.
	I am ____ sorry, teacher. I got up _____ today but there was a traffic jam.
	I __________ believe you. You must have gotten up _______ today.
	No, it is not ______ .
	I think you have been ___ for school _____ . Today is your last warning, ok?
	It won't happen again. Please forgive me.

Lecture F **Episode on offshore!!**

Work in pair. Ask and answer the question.

	When and where were the first submerged oil wells drilled from platforms built on piles in fresh water ?
	They were drilled in the fresh water of the Grand Lake St. Marys in Ohio around 1891.
	When and where were the first submerged oil wells drilled in salt water?
	They were drilled in the portion of the Summerland field extending under the Santa Barbara Channel in California around 1896.
	What well was located on the artificial island in a shallow portion of the Caspian Sea?
	It was the Bibi Eibat well.
	Who drilled the first oil well out of sight of land?
	It was Kerr-Mcgee. oil industries.
	*Make your own!*
	*Make your own!*

OUTRO

Whale Oil Versus The Others

By Samuel T.Pees

The first principal use of whale oil was as an illuminant in lamps and as candle wax. Other uses came in time. In the 1700's it was noted that the burning oil from sperm whales glowed brightly and clearly and did not have a disagreeable odor like the oil from right whales did. The sperm whale was the main whale being sought for its oil when the petroleum industry opened in 1859. The whale fishery, however, was in a declining state and had been so a decade or more before Drake struck petroleum in his drilled well and before general refining of crude oil commenced in Oil Creek Valley and elsewhere.

One would think that there would have been a great competitive clash between whale oil and kerosene from coal and petroleum in the opening years of the 1860's. However, these illuminants did not earnestly join in battle for the U.S. market at that time because the Civil War, beginning in April, 1861, brought the New England whaling fleet to a virtual halt. A large number of the whaling ships were captured and sunk by the Confederacy.

This hazard made an expedition dangerous before the whaling waters were even reached. Nevertheless, sperm whale oil production carried on with its normally expected highs and lows.

Kerosene from petroleum steadily replaced sperm oil as an affordable illuminant in North America and found a great demand in Europe and elsewhere on the globe. Gas manufactured from bituminous coal, water gas and gas made from petroleum distillate came on the scene for illumination and other uses. Manufactured gas lit cities and factories as well as homes.

The next illuminant was natural gas piped to cities and homes from the gas fields. Its use began in 1872 in Titusville, Pennsylvania, via a five mile pipeline from the Newton gas well to the town. Titusville was then an oil capital. Following the sequence of illuminating, natural gas for lighting was eclipsed by electricity.

Through all this, whale fisheries continued to hunt the sperm whales, and a great number of uses for the oil and the other whale products continued to develop. However, refined products from petroleum began to replace some of these other products as well and even whale ambergris, the valuable base for perfumes, was finally replaced by synthetics.

Pops English

The Rose

Some say love, it is a river
that drowns the tender reed.
Some say love, it is a razor
that leaves your soul to bleed.

Some say love, it is a hunger,
an endless aching need.
I say love, it is a flower,
and you its only seed.

It's the heart afraid of breaking
that never learns to dance.
It's the dream afraid of waking
that never takes the chance.
It's the one who won't be taken,
who cannot seem to give,
and the soul afraid of dyin'
that never learns to live.

When the night has been too lonely
and the road has been to long,
and you think that love is only
for the lucky and the strong,
just remember in the winter
far beneath the bitter snows
lies the seed that with the sun's love
in the spring becomes the rose.

Nothing is impossible.
The word itself says I'm possible...

– Audrey Hepburn

UNIT 2

How
Do FPSOs Work?

왜 FPSO가 필요한가?

Lecture A How Do FPSOs Work?

Floating Production Storage and Offloading vessels, or FPSOs, are offshore production facilities that house both processing equipment and storage for produced hydrocarbons. The basic design of most FPSOs encompasses a ship-shaped vessel, with processing equipment, or topsides, aboard the vessel's deck and hydrocarbon storage below in the double hull. After processing, an FPSO stores oil or gas before offloading periodically to shuttle tankers or transmitting processed petroleum via pipelines.

Moored in place by various mooring systems, FPSOs are effective development solutions for both deepwater and ultra-deepwater fields. A central mooring system allows the vessel to rotate freely to best respond to weather conditions, or weather vane, while spread-mooring systems anchor the vessel from various locations on the seafloor. Usually tied to multiple subsea wells, FPSOs gather hydrocarbons from subsea production wells through a series of in-field pipelines. Once tapped by subsea wells, hydrocarbons are transmitted through flow lines to risers, which transport the oil and gas from the seafloor to the vessel's turret and then to the FPSO on the water's surface.

The processing equipment aboard the FPSO is similar to what would be found atop a production platform. Usually built in modules, FPSO production equipment can consist of water separation, gas treatment, oil processing, water injection and gas compression, among others. Hydrocarbons are then transferred to the vessel's double-hull for storage. Crude oil that is stored onboard is frequently transferred to shuttle tankers or ocean barges going ashore, via a loading hose. Loading oil from the stern of the FPSO to the bow of the shuttle tanker is known as tandem loading. Gas is many times transferred to shore via pipeline or re-injected into the field to boost production.

Permanently moored, FPSOs are viable development solutions for a number of different offshore field situations. Because FPSOs can be disconnected from their moorings, these offshore production vessels are optimal for areas that experience adverse weather conditions, such as cyclones and hurricanes. Additionally, because FPSOs can be moved, they are a more economical solution for more marginal fields, in that the vessel can be moved to another development and redeployed once the original

mooring system 계선 계류 장치 any permanent structure to which a vessel may be secured **transmit** 전하다 transfer to another **encompass** 망라하다 아우르다 include in scope **seafloor** 해저 the bottom of a sea or ocean **production platform** 해저 석유 체굴 플랫폼 **offloading** 하역 **weather vane** 풍향계 **injection** 주입 the forceful insertion of a substance under pressure **redeploy** 재배치하다 deploy anew field has been depleted. Also, FPSOs are an optimal choice for development when there are no existing pipelines or infrastructure to transfer production to shore. Adding to the economic advantages of FPSOs, existing tankers are frequently converted into FPSOs.

Used in offshore production since the 1970s, FPSOs have been historically utilized in the North Sea, offshore Brazil, Asia Pacific, the Mediterranean Sea and offshore West Africa. Oil spills do not usually occur from FPSOs, although in the late 1990s the Texaco Captain FPSO spilled approximately 3,900 barrels of oil due to human error. Besides this incident, FPSOs have spilled less than approximately 500 barrels of oil combined.

Besides FPSOs, similar floating systems include Floating Storage and Offloading systems (FSOs), Floating Production Systems (FPSs) and Floating Storage Units (FSUs). Additionally, the world's first FDPSO, or Floating Drilling Production Storage and Offloading vessel, was developed in 2009 for Murphy Oil's Azurite field offshore Republic of Congo. This Azurite FDPSO incorporates deepwater drilling equipment that will help to develop the field and can be removed and reused after all the Azurite production wells have been drilled. Furthermore, the world's first FLNG or Floating Liquid Natural Gas vessel is currently being developed.

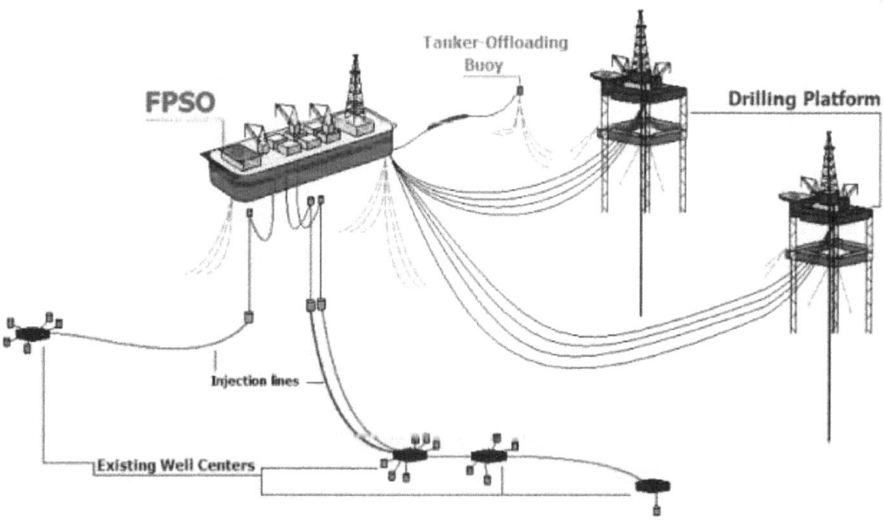

spill 흘리다 the act of allowing a fluid to escape **approximately** 대체로 imprecise but fairly close to correct **combine** 결합하다 합치다 put or add together **furthermore** 더욱이 in addition **besides** 게다가 making an additional point

Lecture B

After you read

1. Fill in the blanks based on this article.

   a. FPSO production equipment can consist of water ________, gas treatment, ________________, water injection and gas compression, among others.

   b. Besides FPSOs, similar floating systems include ______________ ,________and ______________ .

2. Which is not part of the basic design of most FPSOs?

   (a) A ship-shaped vessel

   (b) Processing equipment aboard the vessel's deck

   (c) Hydrocarbon storage below in the double hull

   (d) Central mooring system

3. What happens to the oil and gas stored on board FPSO?

   (a) They got offloaded to shuttle tanker or transmitted to onshore facility via pipeline.

   (b) They got refined on board FPSOs.

   (c) They got transported to onshore facility by FPSOs.

   (d) They got transferred to FSO for long-term storage.

4. Which one of the following is **not** true?

   (a) FPSOs are effective development solutions for both deepwater and ultra-deepwater fields.

   (b) Crude oil that is stored onboard is frequently transferred to shuttle tankers or ocean barges going ashore, via a loading hose.

   (c) FPSOs can' t be disconnected from their moorings, these offshore production vessels are optimal for areas .

Lecture C **Language Focus**

1. Match the words with the definitions, then complete the definition with the given words.

Words	Definitions
mooring system	________ production ________ and offloading vessel
FDPSO	Floating ________ production Storage and ________ vessel
FPSO	any ________ structure to which a vessel may be ________.

offloading floating secured drilling permanent storage

2. Complete the sentences using the following words.

spill combine transmit besides

1) Hydrocarbons are ________ through flow lines to risers.
2) ________ FPSOs, similar floating systems include FSOs, FPSs and FSUs.
3) Oil ________ do not usually occur from FPSOs.
4) The Pacific Ocean is larger than all of the Earth's land area ________ .

3. Complete the sentences using words you' ve studied in this article.

1) Once _____by subsea wells, hydrocarbons are transmitted through flow lines to risers,
   a)tapping b)tapped c) have been tapped d) is tapping
2) The world's first FLNG vessel is currently ________.
   a)being developing b) developed c) developing

Lecture D **Useful Expressions - one~ the other---**

One ~ the other---

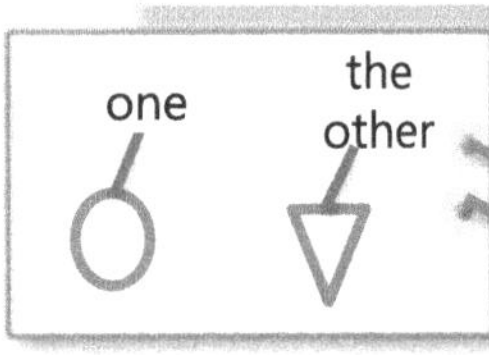

I have two sisters. **One** is tall and **the other** is short.

One ~ the others---

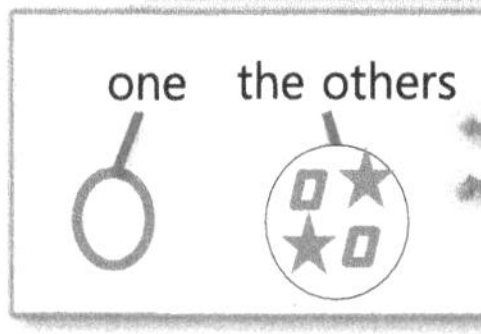

I have five sisters. **One** is tall and **the others** are short.

One ~ another --- the other

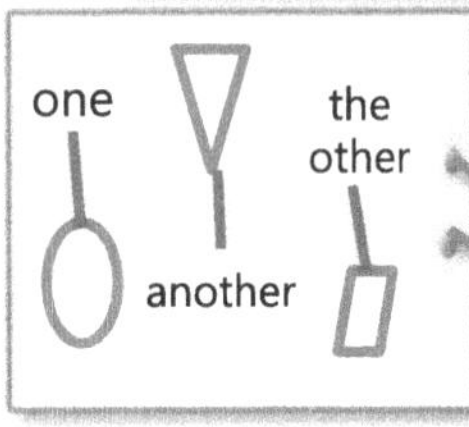

I have three dogs. **One** is yellow, **another** is white, and **the other** is brown .

One ~ another---

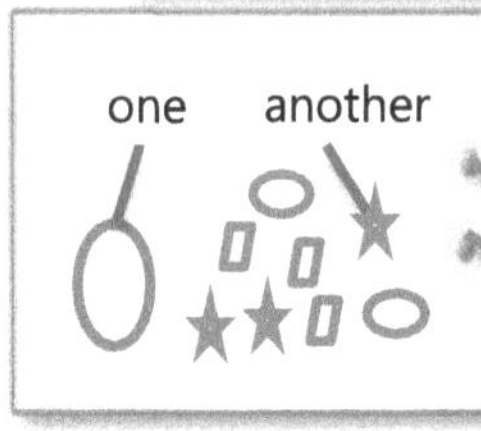

I think this is so big for me. Can you show me **another** t-shirt ?

Some ~ the others---

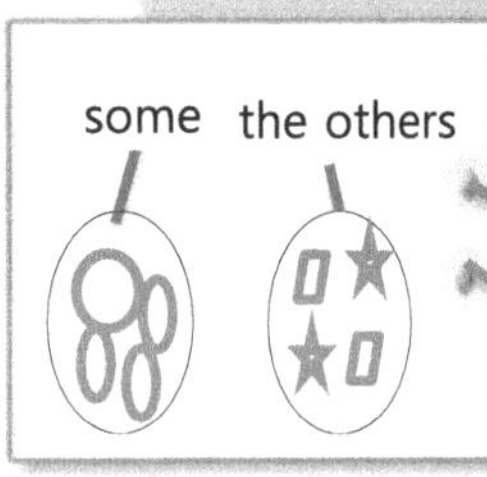

I bought a bunch of flowers. **Some** are roses and **the others** are lilies.

Some ~ others---

There are so many flowers in the garden. **Some** are roses and **others** are lilies.

Put in the blank using expressions above.

1. Of the 30 people, ________ say it is true, but _________ say it is not.
2. Some animals have a backbone, while ___________ don't.
3. There are three people. _____ is a doctor, __________ is a teacher and ____________ is a lawyer.
4. There are several ways to making friends. Being a nice person is ______. Keeping smiling is ___________ .
5. You don't need to listen to him. Let it go in one ear and out _________ .
6. I have two brothers. ___________ is Bill and ____________ is Tom.

Lecture E	Speak Out!!

1. Talk with your partner. What would you say in these situation? Complete the sentences using the expressions you've studied in this unit .

You say

Castle Museum 	A: I heard you've been there before. Can you recommend some places to go? B:There are two places that I'd like to recommend to you. ________ is an old Castle and _____ is a museum.
1 brown dog 4 white dogs 	I have five dogs. ________ is a brown puppy and ________ are white puppies with black ears.
Strawberries Tomatoes 	A: Wow, look at those. There are so many fruits. B: Yes, I really like all kinds of fruits. ________ are strawberries and ________ are tomatoes. These are my favourite ones.

2. Have a conversation with your partner using the words below.
You can use your own ideas.

example 1 red / green/ yellow
A: There are five apples. Can you describe them?
B: <u>One</u> is red, <u>another</u> is green and <u>the others</u> are yellow.

	a. red / green A : B :
	b. Long pants / short pants / shoes A : B :

Lecture F **Episode on offshore!!**

Work in pair. Ask and answer the question.

	What is a FPSO?
	Floating Production Storage and Offloading vessels, or FPSOs, are offshore production facilities that house both processing equipment and storage for produced hydrocarbons
	What can you tell me about the basic design of FPSO?
	The basic design of most FPSOs encompasses a ship-shaped vessel, with processing equipment, or topsides, aboard the vessel's deck and hydrocarbon storage below in the double hull.
	How does crude oil get transferred to the shuttle tankers ?
	Crude oil that is stored onboard is frequently transferred to shuttle tankers or ocean barges going ashore, via a loading hose. Loading oil from the stern of the FPSO to the bow of the shuttle tanker is known as tandem loading. While gas is many times transferred to shore via pipeline or re-injected into the field to boost production.
	Make your own!
	Make your own!

OUTRO

Pipeline Transport

There is some argument as to when the first crude oil pipeline was built. However, some say pipeline transport was pioneered by Vladimir Shukhov and the Branobel company in the late 19th century. Others say oil pipelines originated when the Oil Transport Association first constructed a 2-inch wrought iron pipeline over a 9.7 km track from an oil field in Pennsylvania to a railroad station in Oil Creek, in the 1860s.

Pipelines are generally the most economical way to transport large quantities of oil, refined oil products or natural gas over land. Compared to shipping by railroad, they have lower cost per unit and higher capacity. Although pipelines can be built under the sea, that process is economically and technically demanding, so the majority of oil at sea is transported by tanker ships.

Oil pipelines are made from steel or plastic tubes with inner diameter typically from 4 to 48 inches. Most pipelines are typically buried at a depth of about 3 to 6 feet. To protect pipes from impact, abrasion, and corrosion, a variety of methods are used. These can include wood slats, concrete coating, high-density polyethylene, rock shield, imported sand padding, and padding machines.

The oil is kept in motion by pump stations along the pipeline, and usually flows at speed of about 1 to 6 metres per second. Multi-product pipelines are used to transport two or more different products in sequence in the same pipeline. Usually in multi-product pipelines there is no physical separation between the different products. Some mixing of adjacent products occurs, producing interface, also known in the industry as "transmix." At the receiving facilities this interface is usually absorbed in one of the products based on pre-calculated absorption rates. Alternately, transmix may be diverted and shipped to facilities for separation of the commingled products.

Crude oil contains varying amounts of wax, or paraffin, and in colder climates wax buildup may occur within a pipeline. Often these pipelines are inspected and cleaned using pipeline inspection gauges, pigs, also known as scrapers or Go-devils.

Smart pigs are used to detect anomalies in the pipe such as dents, metal loss caused by corrosion, cracking or other mechanical damage. These devices are launched from pig-launcher stations and travel through the pipeline to be received at any other station down-stream, either cleaning wax deposits and material that may have accumulated inside the line or inspecting and recording the condition of the line.

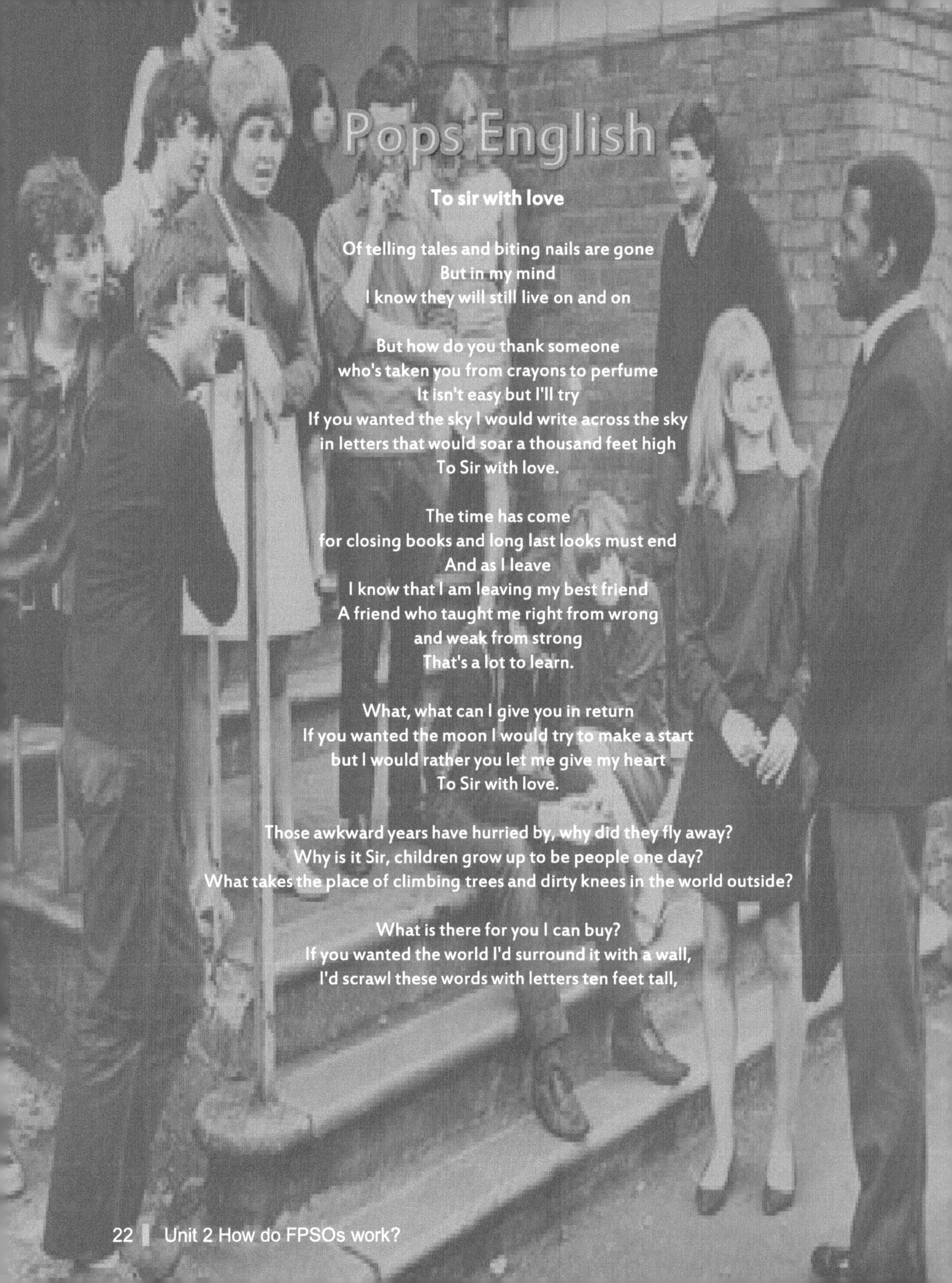

Pops English

To sir with love

Of telling tales and biting nails are gone
But in my mind
I know they will still live on and on

But how do you thank someone
who's taken you from crayons to perfume
It isn't easy but I'll try
If you wanted the sky I would write across the sky
in letters that would soar a thousand feet high
To Sir with love.

The time has come
for closing books and long last looks must end
And as I leave
I know that I am leaving my best friend
A friend who taught me right from wrong
and weak from strong
That's a lot to learn.

What, what can I give you in return
If you wanted the moon I would try to make a start
but I would rather you let me give my heart
To Sir with love.

Those awkward years have hurried by, why did they fly away?
Why is it Sir, children grow up to be people one day?
What takes the place of climbing trees and dirty knees in the world outside?

What is there for you I can buy?
If you wanted the world I'd surround it with a wall,
I'd scrawl these words with letters ten feet tall,

Hope is a waking dream.

- Aristotle

UNIT 3

How Do Semi-submersible Drilling Rigs Work?

반잠수식 시추선은 무엇인가?

Lecture A How Do Semi-submersible Drilling Rigs Work?

Originally conceived as a bottom-supported drilling unit, semi-submersibles eventually found their true calling. Now, semi-subs are the most stable floating rig chosen for harsh conditions because of their ability to withstand rough waters. A semi-submersible is a MODU designed with a platform-type deck that contains drilling equipment and other machinery supported by pontoon-type columns that are submerged into the water. Another type of drilling rig that can drill in ultra-deep waters is drillships and they are capable of holding more equipment; but semisubmersibles are chosen for their stability. The design concept of partially submerging the rig lessens both rolling and pitching on semisubs.

While in transit, semisubs are not lowered into the water. Only during drilling operations semisubs are partially submerged. Because semisubs can float on the top of the water, transporting these rigs from location to location is made easier. Some semisubs are transported via outside vessels, such as tugs or barges, and some have their own propulsion method for transport. Based on the way the rig is submerged in the water, there are two main types of semisubmersibles such as bottle-type semisubs and column-stabilized semisubs. Bottle-type semisubs consist of bottle-shaped hulls below the drilling deck that can be submerged by filling the hulls with water. The first incarnation of this type of drilling rig, bottle-type semisubs originally were conceived as submersible rigs. As a submersible, the bottles below the rig were completely submerged, resting on the ocean floor.

But, as time progressed, naval architects realized that the rig would maintain its stabilization if the bottles were only partially submerged, but be able to drill in deeper waters. Mooring lines are then used to keep the semisub in place, and these anchors are the only connection the rig has with the sea floor. Eventually, these bottle-type rigs were designed to only serve as semisubs. As a semisub, the rig offered exceptional stability for drilling operations, and rolling and pitching from waves and wind was great diminished. In addition to occasional weather threats, such as storms, cyclones or hurricanes, some drilling locations are always harsh with constant rough waters. Being able to drill in deeper and rougher waters, semisubs opened up a new avenue for exploration and development operations

semi-submersible a specialized marine vessel with good stability and seakeeping stable **harsh**= rough 거친 severe **withstand** 견디다 resist or confront with resistance **submerged** 물속의 가라앉은 beneath the surface of the water transit 통과 pass across
propulsion 추진 a propelling force exceptional

A more popular design for semisubmersible rigs is the column-stabilized semisub. Here, two horizontal hulls are connected via cylindrical or rectangular columns to the drilling deck above the water. Smaller diagonal columns are used to support the structure. Submerging this type of semisub is achieved by partially filling the horizontal hulls with water until the rig has submerged to the desired depth. Mooring lines anchor the rig above the well, and dynamic positioning can help to keep the semisub on location, as well.

Semisubmersibles and other mobile offshore facilities are moored in systematic ways, but there are many different designs for various situations. Mooring is similar to multiple anchors, and a number of spread mooring patterns are used to keep the floating rig in place. These mooring spreads are chosen depending on the shape of the vessel being moored and the sea conditions in which it will be moored.

As the wellbore is extremely precise, it is very important that the semisub is kept in position, despite the waves and the winds working to move it about. Furthermore, working in ultra-deepwaters, the drilling riser pipe may span thousands of feet from the bottom of the semisub to the stationary subsea well equipment located on the ocean floor. The drilling equipment is somewhat flexible to overcome slight movements caused by the wind and waves, but the drilling risers must not be bent beyond what it can manage, or they will break. Additionally, dynamic positioning can be used as well as mooring lines to keep the rig in place. Dynamic positioning uses different motors or propulsion units on the vessel to counteract against the motions of the water. Many times, the dynamic positioning system is guided by telemetry signals from beacons on the ocean floor, satellite information and the angular movements of a cable.

horizontal 수평의 parallel to or in the plane of the horizon or a base line **symmetric** 대칭의 having similarity in size, shape, and relative position of corresponding parts **satellite** 위성 man-made equipment that orbits around the earth or the moon **welbore** a hole made by drilling or boring: a borehole **Dynamic positioning** 동적위치유지 position control of ships using thrusters and propellers as actuators to keep the ship at a reference position relative to the seafloor or relative to another ship or platform.

Lecture B **After you read**

1. Why made semi-subs the most stable of any floating rig for harsh conditions?
   (a) Their ability to withstand rough waters
   (b) Their ability to have their own propulsion
   (c) The capability of holding more equipment
   (d) The capability of keeping position by DP system

2. Which design concept can lessen both rolling and pitching on semisubs?
   (a) Partial submerging
   (b) A platform-type deck
   (c) A bottom-supported drilling
   (d) Pontoon-type columns

3. When is semi-submersible partially submerged ?
   (a) during transit
   (b) during drilling
   (c) during positioning
   (d) during offloading

4. Which one of the following is not true?
   (a) The drilling equipment is somewhat flexible to overcome slight movements caused by the wind and waves.
   (b) the drilling risers must not be bent beyond what it can manage, or they will break.
   (c) Many times, the dynamic positioning system is guided by telemetry signals from beacons on the ocean floor, satellite information and the angular movements of a cable.
   (d) Semisubmersibles are moored only by spreading mooring system for various situations.

Lecture C	Language Focus

1. Match the words with the definitions. Then complete the definition with the given worlds.

satellite	●	a specialized marine vessel with good ___________ and seakeeping stable
semi-submersible	●	___________ equipment that _______ around the earth or the moon
submerged	●	_______________ the surface of the water

orbits   stability   man-made   beneath

2. Complete the sentences using the following words.

tugs or barges   transit   keep   propulsion   lowered

1) While in ___________, semisubs are not ___________ into the water.
2) Some semis are transported via outside vessels, such as _____________, and some have their own _____________ method for transport.
3) Mooring lines anchor the rig above the well, and dynamic positioning can help to _____ the semisub on location, as well.

3. Answer these questions using new vocabulary you studied in this article .

1) This is _____________ by pontoon-type columns that are submerged into the water.
   a) having supported b) supports c) supported d) is supporting
2) The drilling equipment is __________ flexible to overcome slight movements.
   a) somewhere b) something c) somehow d) somewhat

Lecture D Useful Expressions – as soon as

A. as+ adjective/adverb + as

as+ adjective/adverb + as	not as[so]+ adjective/adverb + as
▪ Her hair is **as black as** coal.	▪ Mrs. Smith is not **as young as** me. =I am younger than Mrs. Smith.
▪ He runs **as quickly as** a cheetah. ▪ He knows the world **as little as** a newborn baby.	▪ His second novel is not **so popular as** the first one. =His first novel was more popular than the second one

B. as +[many/much/few/little] +(noun)+ adjective/adverb + as

I ate **as many** potato chips as you did.

Please complete this form **as much** information as possible.

C. half[twice, three times] + as +adjective/adverb + as

I have only **half as** many toys as he does.

My dad is **twice as** tall as my little brother.

Her house is **four times as** big as my house.

D. as +adjective/adverb + as + S+V/possible

I am sorry for late, but I got here **as fast as** I could. (=as fast as possible)

There is a plenty of food. You can have **as much as** you want. (=as much as possible)

Please write me back **as soon as** you can. (= as soon as possible)

Tell me about your plan **as specifically as** you can. (=as specifically as possible)

I jumped **as high as** to reach the ceiling. (=as high as possible)

Lecture E	**Speak Out!!**

1. Talk with your partner. What would you say in these situation? Complete the sentences using the expressions you've studied in this unit .

Situation	You say
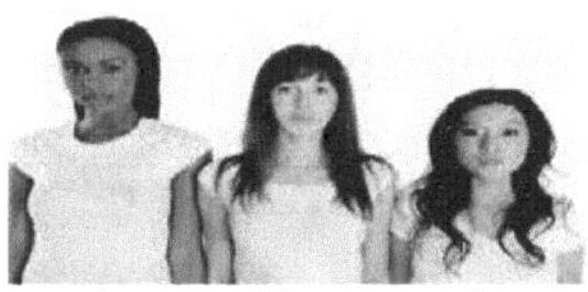 Ann Hanna Jay	Hanna is not ________________ Ann. (tall) Ann is not ________________ Jay. (short) Jay is not ______________ Hanna. (tall)
	He was poor while growing up. However, now he runs a big business company. His house is _______ as _____ as used to be (X10/ big).
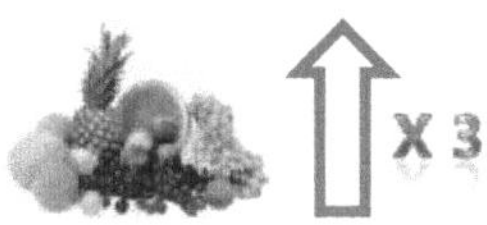	Fruits are very expensive now. The price is ___________ as __________ usual. (X3/ high).

2. Have a conversation with your partner using the words below and practice twice. You can use your own ideas.

Ex.

 I watched the movie, *The Love 2* yesterday.

 Really? How did you like it?

 I think it was as good as The Love 1.

 What makes you think so?

 It was so nice, I really enjoyed it.

 I see. I want to watch that movie, too.

A.

- A top singer
- New album / come out
- really good.

1st Album 2nd Album

B.

- See/ TV star
- as good as on TV

Lecture F Episode on offshore!!

Work in pair. Ask and answer the question.

	Why is semi-subs known to be the most stable of any floating rig for harsh conditions?
	The semi-subs are the most stable of any floating rig, many times chosen for harsh conditions because of their ability to withstand rough waters.
	What is a semi-submersible?
	A semi-submersible is a MODU designed with a platform-type deck that contains drilling equipment and other machinery supported by pontoon-type columns that are submerged into the water.
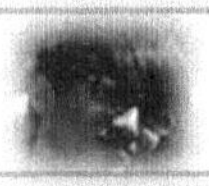	What is a bottle -type semi-submersible?
	A bottle-type semisubs consist of bottle-shaped hulls below the drilling deck that can be submerged by filling the hulls with water.
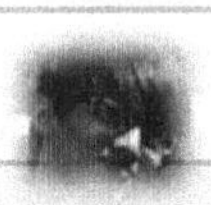	What is a column-stabilized semi-subsible?
	In a bottle -type semi-submersible, two horizontal hulls are connected via cylindrical or rectangular columns to the drilling deck above the water. Smaller diagonal columns are used to support the structure.
	*Make your own!*
	*Make your own!*

What is barrel?

The wooden oil barrel of the late 19th century is different from the modern-day 55-gallon steel drum (known as 210/200-litre/kg outside the USA; sometimes the 44-gallon drum in Britain). The 42-US-gallon oil barrel is a unit of measure, and is no longer a physical container used to transport crude oil, as most petroleum is moved in pipelines or oil tankers.

In the United States, the 42-US-gallon size of barrel as a unit of measure is largely confined to the oil industry, while different sizes of barrel are used in other industries. Nearly all other countries use the metric system. Many oil-producing countries still use the American oil barrel.

The measurement originated in the early Pennsylvania oil fields. In the early 1860s, when oil production began, there was no standard container for oil, so oil and petroleum products were stored and transported in barrels of different shapes and sizes.

Some of these barrels would originally have been used for other products, such as beer, fish, or whisky. Both the 42-US-gallon barrels based on the old English wine measure, the tierce (159 litres) and the 40-US-gallon (151.4-litre) whiskey barrels were used. 45-gallon barrels were also in common use. The 40-gallon whiskey barrel was the most common size used by early oil producers, since they were readily available at the time.

The origins of the 42-gallon oil barrel are obscure, but some historical documents indicate that around 1866, early oil producers in Pennsylvania came to the conclusion that shipping oil in a variety of different containers was causing buyer distrust.

They decided they needed a standard unit of measure to convince buyers that they were getting a fair volume for their money. They agreed to base this measure on the more-or-less standard 40-gallon whiskey barrel, but, as an additional way of assuring buyer confidence, they added an additional two gallons to ensure that any measurement errors would always be in the buyer's favor.

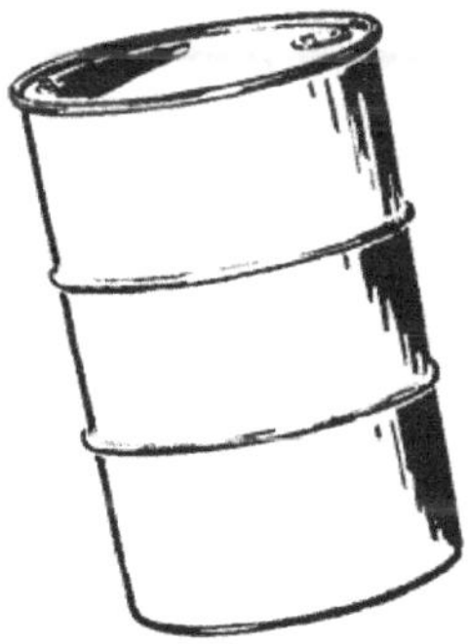

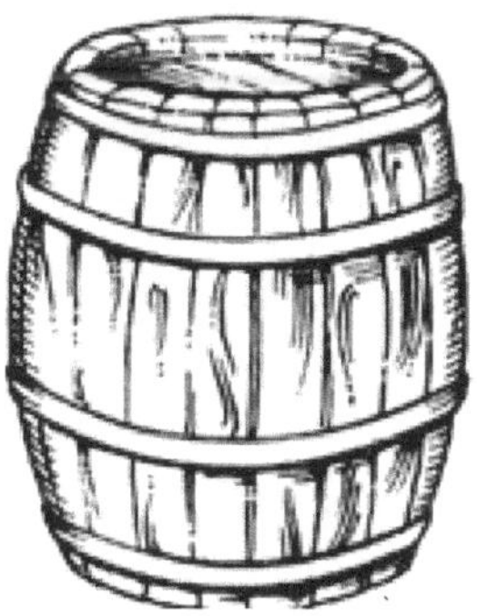

Pops English

My Heart Will Go On

Every night in my dreams I see you, I feel you,
That is how I know you go on,
Far across the distance and spaces between us
You have come to show you go on

Near, far, wherever you are,
I believe that the heart does go on.
Once more, you open the door
And you're here in my heart,
And my heart will go on and on

Love can touch us one time and last for a lifetime,
And never let go till we're gone.
Love was when I loved you, one true time I hold to
In my life we'll always go on.

Near, far, wherever you are,
I believe that the heart does go on.
Once more, you open the door
And you're here in my heart,
And my heart will go on and on.

You're here, there's nothing I fear,
And I know that my heart will go on.
We'll stay forever this way,
You are safe in my heart,
And my heart will go on and on.

Innovation distinguishes between
a leader and a follower.
- Steve Jobs

UNIT 4

What

is a Jack Up Drilling Rig?

Jack up Drilling rig는 무엇인가?

Lecture A

What Is a Jack-up Drilling Rig?

A jack-up rig or a self-elevating unit is a type of mobile platform that consists of a buoyant hull fitted with a number of movable legs, capable of raising its hull over the surface of sea. The buoyant hull enables transportation of the unit and all attached machinery to a desired location. Once on location the hull is raised to the required elevation above the sea surface on its legs supported by the sea bed. The legs of such units may be designed to penetrate the sea bed, may be fitted with enlarged sections or footings, or may be attached to a bottom mat.

Since the first jackup was built in 1954, jackups have become the most popular type of mobile offshore drilling unit (MODU) for offshore exploration and development purposes. In fact, there are hundreds of jackups stationed around the world performing drilling and workover operations in just about every environment. Jack-ups are one of two types of bottom-supported MODUs. Jackup rigs rest on the sea floor rather than float. The premise of a jackup rig is that it is self-elevating; here, the legs are stationed on ocean floor and the drilling equipment is jacked up above the water's surface.

Providing a very stable drilling environment, in comparison to other types of offshore drilling rigs, jackups can drill in waters up to 350 feet deep. Once drilling is required in waters that are deeper than the capabilities of a jackup, semisubmersibles and drillships become a more logical choice for exploration and development operations. When their legs are not deployed, jackups float, which makes these types of MODUs quite easily transported from one drilling location to another. While some are capable of self-propulsion and do not need an outside source for movement, most jackups are transported via tug boats or submersible barges. While towing is easily performed, barges are the transportation of choice when the jackup needs to be moved quickly or over a longer distance.

Whether a jackup design involves three or four legs, there are two main types of legs that are used on jackups. The first incorporates open-truss legs into the design that resemble electrical towers. Open-truss legs are made of tubular steel sections that are crisscrossed, making them strong, but lightweight. Jackups can also have columnar legs made of huge steel tubes. While columnar legs are less expensive than open-truss legs to fabricate, they are less stable and cannot adapt to stresses in the water as well as open-truss legs. For this reason, columnar-legged jackups are not used in waters that measure more than 250 feet deep. In addition to their legs, jackups are supported by two

premise 전제 a statement that is assumed to be true and from which a conclusion can be drawn **open-truss** 열린지붕 a framework of beams forming a rigid structure that supports a roof or bridge or other structure **resemble** 닮다 be similar **spud can** the foot on a leg of a 'Jack-up' type oil platform, designed to spread the load and prevent over penetration of the leg into the sea bed **rest on** 놓여있다 rest on for support **incorporates** 통합하다 formed or united into a whole **tubular** 관모양의 constituting a tube different systems of stabilization. Jackup legs are supported on the sea floor via either mats or spud cans. Usually used on independent-legged jackups, spud cans are cylindrically shaped steel shoes with pointed ends, similar to a cleat. Spud cans are attached to the bottom of each leg, and the spike in the can is driven into the ocean floor, adding stability to the rig during operations.

Additionally, there are two types of elevating devices. Once the jackup is on location, the legs are lowered to the ocean's floor and the rig hull and drilling equipment is elevated well above the water's surface and away from any potential waves. The first type of elevating device uses hydraulic cylinders equipped with moving and stationary pins. These cylinders extend and retract to climb up and down the legs of the jackup. The other type of elevating device employs a rack and two pinion gears that are turned to move the legs up and down. Whichever type of leg that is used in the design, the legs of the jackup rise up through holes in the hull of the drilling rig. Here, a deck is used to support the drilling derrick and other equipment.

There are two ways to mount the drilling equipment on the hull. The most popular design for drilling equipment is a cantilevered jackup. Here, the drilling derrick is mounted on an arm that extends outward from the drilling deck. With a cantilevered jackup, drilling can be performed through existing platforms, as well as without them. Because of the range of motion that the cantilever provides, most jackups built in the last 10 years have been cantilevered jackups. The other type of jackup is the slot-type jackup, also known as a keyway jackup. Drilling slot jackups are built with an opening in the drilling deck, and the derrick is positioned over it. While exploration wells can be drilled with drilling slot jackups, the rig can also be jacked up over another smaller facility, drilling through its hull.

cleat 쐐기, 밧줄걸이 **attach** 붙이다 **elevate** 올리다 raise from a lower to a higher position **device** 장치 an instrumentality invented for a particular purpose **hydraulic** 수압의 moved or operated or effected by liquid **stationary** 정지된 **retract** 원위치하다, 움추려들다 **employ** 사용하다 ,두다 **cantilever** 외팔보 projecting horizontal beam fixed at one end only **mount** 세우다 **range** 범위 slot-type jackup **derrick** 유정탑 a framework erected over an oil well to allow drill tubes to be raised and lowered

Lecture B

After you read

1. When was the first jackup built?
   (a) In 1954
   (b) In 1945
   (c) In 1854
   (d) In 1845
2. How deep can jackups drill in waters?
   (a) Up to 35 meters deep
   (b) Up to 350 feet deep
   (c) Up to 3,500 feet deep
   (d) Up to 35 feet deep

3. There are two ways to mount the drilling equipment on the hull. What is the most popular design for drilling equipment in jackup?
   (a) A cantilevered jackup
   (b) A slot type jackup

4. Which one of the following is not true?
   (a) Once drilling is required in waters that are deeper than the capabilities of a jackup, semisubmersibles and drillships become a more logical choice for exploration and development operations.
   (b) Open-truss legs are made of tubular steel sections that are crisscrossed, making them strong, but lightweight.
   (c) Columnar legs are less expensive than open-truss legs to fabricate and they are more stable to adapt to stresses in the water.
   (d) Once the jackup is on location, the legs are lowered to the ocean's floor and the rig hull and drilling equipment is elevated well above the water's surface.

Lecture c	Language Focus

1. Match the words with the definitions. then complete the definition with the given worlds.

Words	Definitions
derrick	the ____ on a leg of a 'Jack-up' type oil platform
cantilever	a framework _____ over an oil well to allow drill tubes to be raised and lowered
spud can	projecting __________beam fixed at one end only

horizontal   foot   erected

2. Complete the sentences using the following words.

retract   hydraulic   attach

1) The first device uses __________cylinders equipped with moving and stationary pins.
2) These cylinders extend and _________ to climb up and down the legs of the jackup.
3) Spud cans are _______________ to the bottom of each leg.

3. Complete the sentences using the following words.

1) There are two ways to _______the drilling equipment on the hull
   a)be mounted   b) mount   c) mounting   d) be mounting
2) The other type of jackup is the slot-type jackup, also ____ as a keyway jackup.
   a)to be known   b)to know   c) known   d) knowing

Lecture D Useful Expressions – the deepest thing

A. the deepest/ the most famous, etc.)

the +

	cheapest hottest, etc
most	famous dangerous, etc

\+

in	with places
of	with a period / plural objects
that	S + V

Example

This platform is **the deepest thing** in the world.
My family is **the most important** thing in my life.
Yesterday was **the hottest** day of the year. This necklace is **the most expensive** jewelry in this shop.
A volunteer work **is the most valuable** experience I've ever had.

B. More expression about superlatives

the superlative + in / of / that~	A love is the greatest word in the world.
=No (other) ~ comparative +than	=No (other) word in the world is greater than a love.
=No (other) ~ as[so] + adjective +as	=No (other) word in the world is as great as a love.
=There is nothing ~ comparative + than	=There is nothing greater than a love in the world.
=comparative + than any other + singular	=A love is greater than any other word in the world.
=comparative + than all the other + plural	=A love is greater than all the other words in the world.

Lecture E Speak Out!!

1. Talk with your partner. What would you say in these situation? Complete the sentences using the expressions you've studied in this unit.

	Your situation	You say
	a. You are outside. You are sweating like a pig.	Oh, God. I think today is ________________ in this summer. (hot/ day)
	b. Your boyfriend has just proposed now.	Oh my God, This is ______________ of my life! (happy/day)
	c. You are drinking really good coffee with a friend.	That is ______________________ I've ever tasted. (good/ coffee)
	d. You are a mother of this baby boy.	He is ________________________ in my life. (important/person)

2. Have a conversation with your partner based on the following example using Lecture D-B section and practice twice. You can use your own ideas.

Ex.

 Have you heard that? She is coming here.

 who are you talking about?

 I mean your favourite movie star.

 Oh my God, really? I like her very much. I think she is the most attractive actress in Korea.

1. Mt. Everest is the highest mountain in the world.
   = No mountain~
   = There is nothing ~
   = higher than any ~

2. This is the most expensive necklace in our shop.
   =No necklace ~
   =There is nothing~
   = ~ than any ~

Lecture F Episode on offshore!!

Work in pair. Ask and answer the question.

	What is a jack-up?
	A jack-up rig or a self-elevating unit is a type of mobile platform that consists of a buoyant hull fitted with a number of movable legs, capable of raising its hull over the surface of sea.
	How does a jack-up work?
	The buoyant hull enables transportation of the unit and all attached machinery to a desired location. Once on location the hull is raised to the required elevation above the sea surface on its legs supported by the sea bed.
	What are two types of elevating devices of the jack-ups?
	The first type of elevating device uses hydraulic cylinders equipped with moving and stationary pins. The other type of elevating device employs a rack and two pinion gears that are turned to move the legs up and down.
	Make your own!
	Make your own!

Burning Ice

Crude oil, coal and natural gas are not the only fossil fuels hidden deep below the surface of the Earth. Right around the globe, enormous quantities of methane hydrates can be found. This white combustible ice consists of water and methane gas. If thawed in a controlled fashion, many billions of tons of methane could be obtained from it.

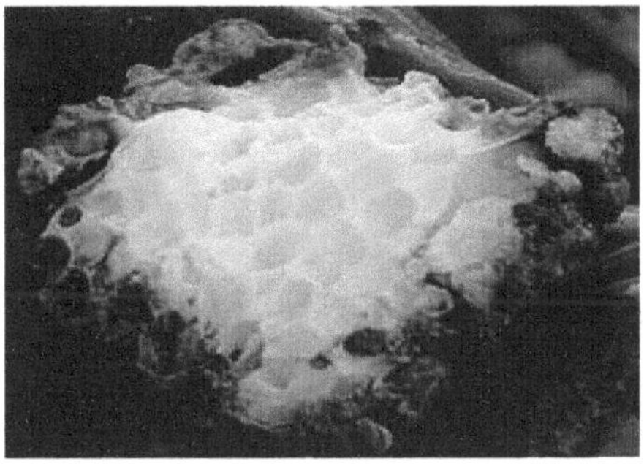

Based on current estimates, known reserves comprise roughly 12,000 billion tons of methane hydrates. Substantial deposits are located on continental shelves in the Pacific around Japan, off the Pacific coastline of America and to the North of Canada and Siberia.

These deposits are located deep down in the seabed and locked into land-based permafrost. Methane hydrates cannot form solid ice crystals until the pressure exceeds 20 bar and temperature fall below four degrees Celsius. The ice evaporates at lower pressure or higher temperature, and in the process, releases 164 times its own volume in the form of gas.

In Japan, Russia, USA and Germany, intensive efforts are under way to find ways to exploit these reserves. However, this methane ice, which has formed over millions of years, would have a devastating impact on our climate. It locks in about twice the volume of carbon as the sum total of all reserves of crude oil, natural gas and coal combined.

As methane burns, this carbon is converted into the greenhouse gas carbon dioxide. On the other hand, it can be used efficiently in modern gas-fired power stations to generate electricity. At this time, researchers are working on the possibility of separating carbon dioxide from the combustion gases of future power stations, and pumping it separately into subterranean storage facilities.

If methane hydrates were ever extracted from continental shelves without appropriate precautionary measures being taken, the shelves on the coasts could become unstable, gigantic landmasses could start to slip and that in turn could trigger tsunamis.

Experts are taking these risks seriously and are only experimenting on the extraction of methane hydrates in flat storage locations far from continental shelves.

Pops English

In dream

A candy colored clown
they call the sandman
tiptoes to my room every night
just to sprinkle stardust
and to whisper
"Go to sleep, everything is alright"
I close my eyes
then I drift away
into the magic night

I softly say a silent prayer
like dreamers do
then I fall asleep to dream
my dreams of you
In dreams I walk with you.
In dreams I talk to you.
in dreams you're mine
All of the time, we're together
in dreams, in dreams

But just before the dawn
I awake and find you're gone.
I can't help it,
I can't help it if I cry
I remember that
you said "Goodbye."
It's too bad that all these things
can only happen in my dreams.
only in dreams.
in beautiful Dreams.

Everything in your world is
created by what you think.
- Oprah Winfrey

UNIT 4

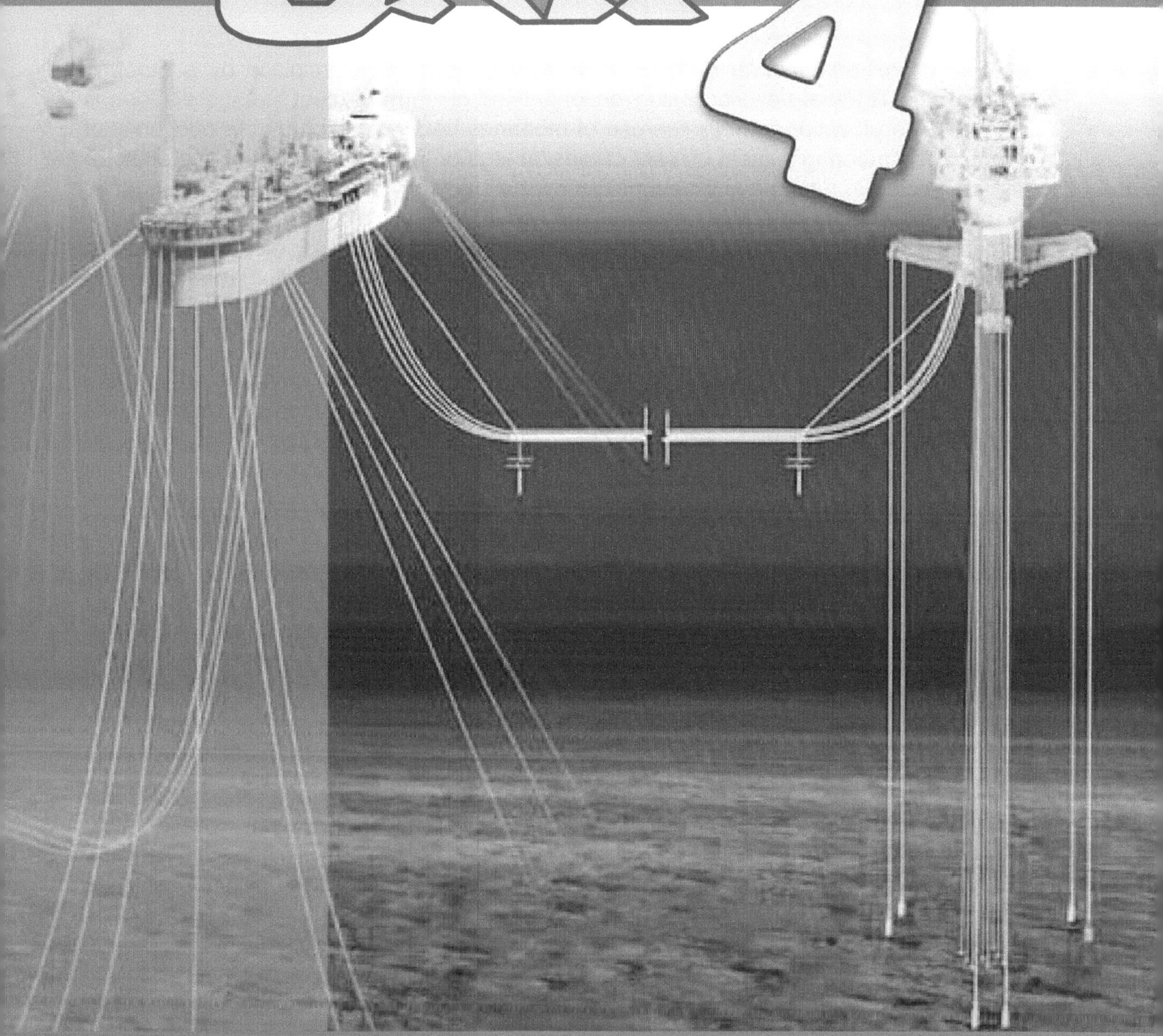

What is a TLP?

TLP는 무엇인가?

Lecture A What is a TLP?

A Tension Leg Platform (TLP) is a buoyant platform held in place by a mooring system. The TLP's are similar to conventional fixed platforms except that the platform is maintained on location through the use of moorings held in tension by the buoyancy of the hull. The mooring system is a set of tension legs or tendons attached to the platform and connected to a template or foundation on the seafloor. The template is held in place by piles driven into the seafloor. This method dampens the vertical motions of the platform, but allows for horizontal movements. The topside facilities (processing facilities, pipelines, and surface trees) of the TLP and most of the daily operations are the same as for a conventional platform.

The foundation is the link between the seafloor and the TLP. Most foundations are templates laid on the seafloor, then secured by concrete or steel piles driven into the seafloor by use of a hydraulic hammer, but other designs can be used such as a gravity foundation. The foundations are built onshore and towed to the site. As many as 16 concrete piles with dimensions of 100 ft in diameter and 400 ft long are used, which is one for each tendon.

The hull is a buoyant structure that supports the deck section of the platform and its drilling and production equipment. A typical hull has four air-filled columns supported by pontoons, similar to a semisubmersible drilling vessel. The deck for the surface facilities rests on the hull. The buoyancy of the hull exceeds the weight of the platform, requiring taut moorings or "tension legs" to secure the structure to the seafloor. The columns in the hull range up to 100 ft in diameter and up to 360 ft in height. the overall hull measurements will depend on the size of the columns and the size of the platform. Modules are units that make up the surface facilities on the deck section of the platform.

Early in TLP development, industry discovered that it is cost effective to build the surface facility in separate units modules, assemble them at shallow inshore location, and then tow them to the site. The modules that are part of a typical TLP include the wellbay, power, process, quarters, and drilling. They are secured to the deck, which is attached to the hull. The typical surface facility will be 65,000 sq ft. The living quarters house up to 100 people, depending on the type and scope of activity being performed. Process capacity ranges up to 150,000 BPD oil and 400 MMscfd gas. A typical drilling rig located on a larger TLP would have a 1.5 million-pound pull derrick, a 2,000-hp top-drive derrick, and three 2,200-hp pumps.

conventional 진부한 , 전통적인 following accepted customs and proprieties **measurement** 측정 **dimension** 치수 **diameter** 지름 **dampen** 축축하게 하다, 꺾다 **assemble** 모으다 taut 팽팽한 **BPD** barrel per day **MMscfd (Million standard cubic feet per day)** a unit of measurement for liquids and gases that is predominantly used in the United States.

A template provides a frame on the seafloor in which to insert either conductors or piles. Not all TLP's use templates; if used, they are typically the first equipment installed at the site. There are several types of templates that may be used in conjunction with a TLP to support drilling, foundation integrity, or the integration of the two. Drilling templates provide a guide for locating and drilling wells.

They may also be a base for the tie-in of flowlines from satellite wells or for export pipelines and their risers. Foundation templates may be one single piece or separate pieces for each corner. The foundation piles are driven through the foundation template. An integrated template is a single piece that contains all drilling support, anchors the tendons, and locates and guides the foundation piles. Separate templates allow each part to be installed individually. They also use smaller pieces that weigh less and are easier to install. The drilling template can be installed and drilling can begin while the foundation template is being designed and built.

Tension Legs (tendons) are tubulars that secure the hull to the foundation. Tendons are typically steel tubes with dimensions of 2-3 ft in diameter with up to 3 inches of wall thickness, the length depending on water depth. A typical TLP would be installed with as many as 16 tendons. A production riser conveys produced fluids from the well to the TLP surface production facilities. An example riser system for a TLP could be either a single-bore or dual-bore (concentric pipe) arrangement. The dual-bore riser would consist of a 21-inch, low pressure (e.g., 3,000 psi) marine riser that serves as an environmental barrier, and an 11 inch inner pipe casing that is rated for high pressures (e.g., 10,000 psi).

Installation of a TLP is done in stages. Often the design work on one section of the TLP is being done while another part is being installed. For example, the wells will often be predrilled while the TLP is being designed and constructed.

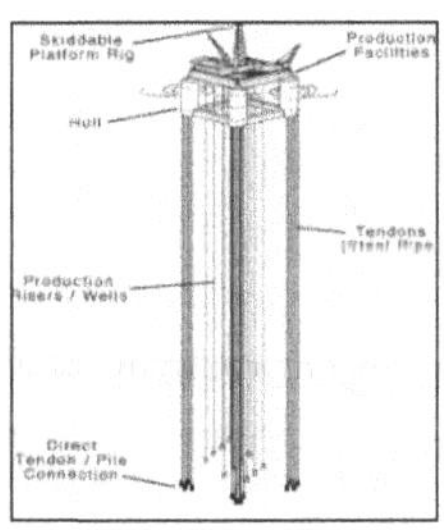

tubular 관모양의 constituting a tube **in conjunction with** 결합하여, 연계하여 **psi** 프사이 압력의 단위. 1평방 인치 당의 파운드(중량). 1in=2.5400㎝, $1in^2$=6.4516㎝, 1lb=453.592g. **secure** 안정한 free from danger or risk **tendon** 철근 **concentric** 집중적인 having a common center **individually** 각각 apart from others **bore** 시추공 **thickness** 두께

Lecture B　　After you read

1. Unlike conventional fixed platforms, how is the TLP platform maintained on location?
   (a) Through the use of moorings held in tension
   (b) By dynamic position
   (c) Through the use of anchoring
   (d) By spread mooring

2. Please complete the sentence.
   (a) A Tension Leg Platform is a ______________ held in place by a mooring system up to 35 meters deep
   (b) The mooring system is a set of _________ or ________ attached to the platform and connected to a template or foundation on the seafloor.
   (c) This method dampens the ________ motions of the platform, but allows for __________ movements.
   (d) The columns in the hull range up to _____ ft in diameter and up to ______ ft in height

3. Which one of the following is not true?
   (a) A template provides a frame on the seafloor in which to insert either conductors or piles.
   (b) Drilling templates provide a guide for locating and drilling wells; they may also be a base for the tie-in of flowlines from satellite wells or for export pipelines and their risers.
   (c) A production riser conveys produced fluids from the well to the TLP surface production facilities
   (d) Tension Legs are steel columns that secure the hull to the foundation.

Lecture c **Language Focus**

1. Match the words with the definitions, then complete the definition with the given worlds.

Word	Definition
MMscfd	tubulars that __________ the hull to the foundation
TLP	a unit of __________for liquids and gases that is ____________used in the United States
Tendon	A __________ platform held in place by a mooring system

measurement secure predominantly buoyant

2. Complete the sentences using the following words.

Dampen in conjunction with conventional

1) The ______________way of looking for a lost pet is to post up fliers and signs.
2) This method ____________the vertical motions of the platform.
3) All members of the association will receive a free copy ____________ their membership fees.

3. Complete the sentences using the following words.

1) Separate templates allow each part to be installed ___________.
   a) individual b) individually c) individualistic d) individualist

2) of a TLP is done in stages.
   a) installation b) install c) Installable

Lecture D Useful Expressions – Each/every

A. Each and Every

Each = 

Every= 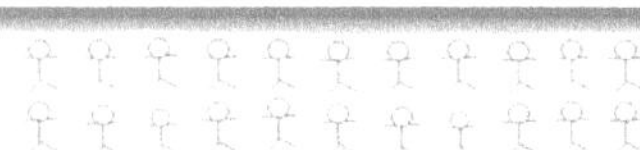

Each student has a unique personality.

Each + singular nouns + verb

Every student has to study hard .

Each + singular nouns + verb

Often it is possible to use each or every. However, sometimes it is not possible.

-In a baseball game, **each** team has nine players. (*not* every)

-How often do you use a computer? **Every** day. (*not* Each day)

-There is a bus **every** 10 minutes. (*not* each 10 minutes)

B. Compare the structures with each and every.

Each can be used alone, or each one.

- **Each** (=each room) is different.

= **Each** one is different.

You can say each of (the /these/them, etc):

- **Each** of the rooms is different.
- **Each** of these rooms is different.
- **Each** of them is different.

Every can't be used alone, but you can say every one.

- Have you bought all these jewelry?
- Yes, **every** one.

You can say every one of (but not every of)

- I've bought **every** one of these jewelry.

=I've bought every one of them.

C. Put in each / every and proper verbs.

1. The book is divided into four parts, and _______ of these _____ three units. (have)
2. _______ parent _________ about their children. (worry)
3. Kate had read _________ book in the library.

Lecture E　　Speak Out!!

A. Talk with your partner. What would you say in these situation? Complete the sentences using the expressions you've studied in this unit.

	1. She has read ____________ in the library. (book)		2. My science team is working well. We've won ____________ this year. (contest)
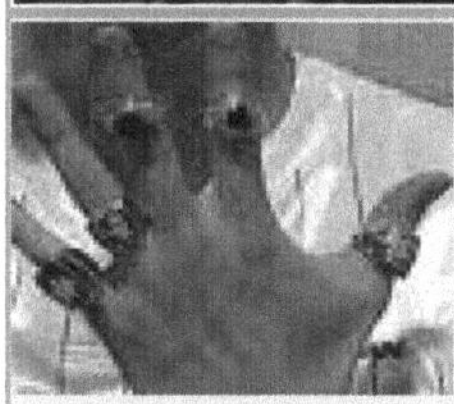	3. She is wearing five rings-one on ______________ . (finger)		4. ____________ in the concert hall was taken. (seat)
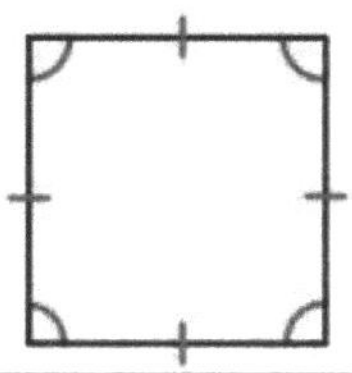	5. ___________ of a square is the same length. (side)		There are interesting apartments in that building. ________ has a unique balcony.

B. Have a conversation with your partner based on the following example using Lecture D-B section and practice twice. You can use your own ideas.

Ex.

	Have you ever been to London?		Of course, I've been there.
	What did you like about London?		Hmm, every building in London was so beautiful and magnificent.
	and, then?		Also, each one had a unique design . It was amazing.

1. every dog /adorable each of them / characteristic		2. Every member/ pay for the entrance fee Each member/ wear his or hers name tag.	

Lecture F Episode on offshore!!

Work in pair. Ask and answer the question.

	What is a TLP?
	A Tension Leg Platform (TLP) is a buoyant platform held in place by a mooring system. The TLP's are similar to conventional fixed platforms except that the platform is maintained on location through the use of moorings held in tension by the buoyancy of the hull.
	How can the hull of TLP be described?
	The hull is a buoyant structure that supports the deck section of the platform and its drilling and production equipment. A typical hull has four air-filled columns supported by pontoons, similar to a semisubmersible drilling vessel.
	How do the tendons work?
	Tension Legs (tendons) are tubulars that secure the hull to the foundation. Tendons are typically steel tubes with dimensions of 2-3 ft in diameter with up to 3 inches of wall thickness, the length depending on water depth. A typical TLP would be installed with as many as 16 tendons.
	*Make your own!*
	*Make your own!*

How the seas were named (1)

Aegean Sea
Named after the Athenian King Aegeus of Greek mythology. His son Theseus promised to hoist a white sail on his voyage home from Crete as a sign that he was still alive. When Theseus neglected to do so, Aegeus thought he had been killed and committed suicide by throwing himself in the sea that came to be named after him

Atlantic Ocean
Takes its name either from the lost continent of Atlantis or more probably, from the Atlas mountains in northwest Africa, which ancients believed overlooked the entire ocean.

Baltic Sea
Its name derives from either Lithuanian baltas, 'white', or from the Scandinavian balta, 'strait'. In the latter case, the Baltic is the sea of straits.

Black Sea
Known as the Euxine or 'The Hospitable'; to the Greeks, even though it was noted for its storms and rocky shores, the Greeks believed it unwise to give derogatory name to any natural force, whether good or bad. The Turks later named is the Black Sea because they were terrified of its dangers and large stretches of open water.

Caribbean Sea
Named after the Caribe tribe of people at the South American coast, first encountered by Columbus. The Caribs gave the name to Columbus as Canibales, a dialectical form of the Caribes, and because it was said that some of these fierce bold warriors ate human flesh, their name, in the form of cannibal, also became to be synonym for man-eaters. Still another word is derived from the Caribs or Canibales is Shakespear's Caliban, Prospero's half human slave in the 'The Tempest' and hence any bestial, degraded man. Caliban was probably formed either by metathesis from the Spanish cannibal or as an variant of Cariban, a Carib.

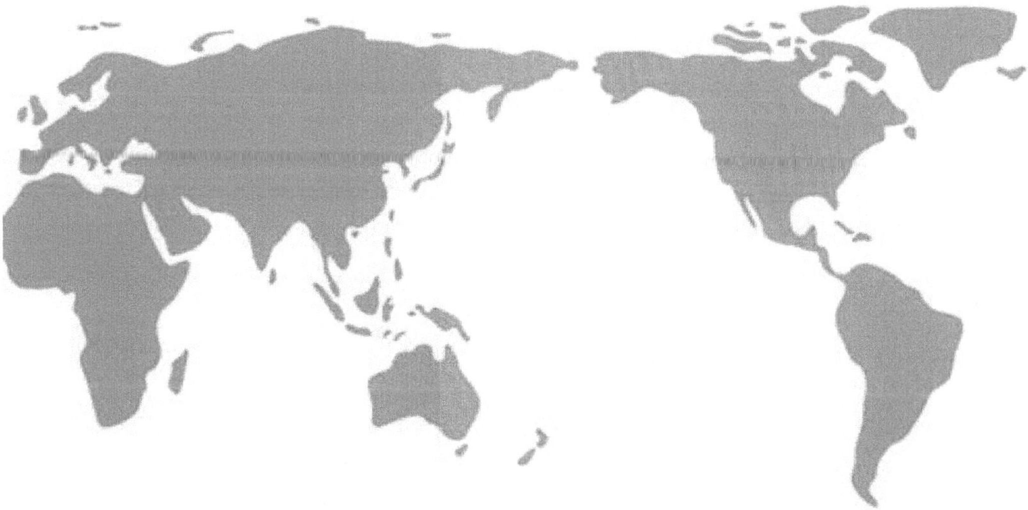

Pops English

Just Called To Say I Love You

No New Year's Day to celebrate
No chocolate covered candy hearts to give away
No first of spring   No SONG to spring   No SONG to sing
In fact here's just another ordinary day
No April rain   No flower bloom
No wedding Saturday within the month of June
But what it is,is something true
Made up of these three words
That I must say to you

I just called to say I love you
I just called to say how much I care
I just called to say I love you
And I mean it from the bottom of my heart

No summer's high   No warm July
No harvest moon to light one tender August night
No autumn breeze   No falling leaves
Not even time for birds to fly to southem skies
No Libra sun   No Halloween
No giving thanks to all the Christmas joy you bring
But what it is though old so new
To fall your heart like no three words could ever do

I just called to say I love you
I just called to say how much I care
I just called to say I love you
And I mean it from the bottom of my heart

I just called to say I love you
I just called to say how much I care
I just called to say I love you
And I mean it from the bottom of my heart

I just called to say I love you
I just called to say how much I care
I just called to say I love you
And I mean it from the bottom of my heart

Try not to become a man of success but rather to become a man of value.

- Albert Einstein

UNIT 6

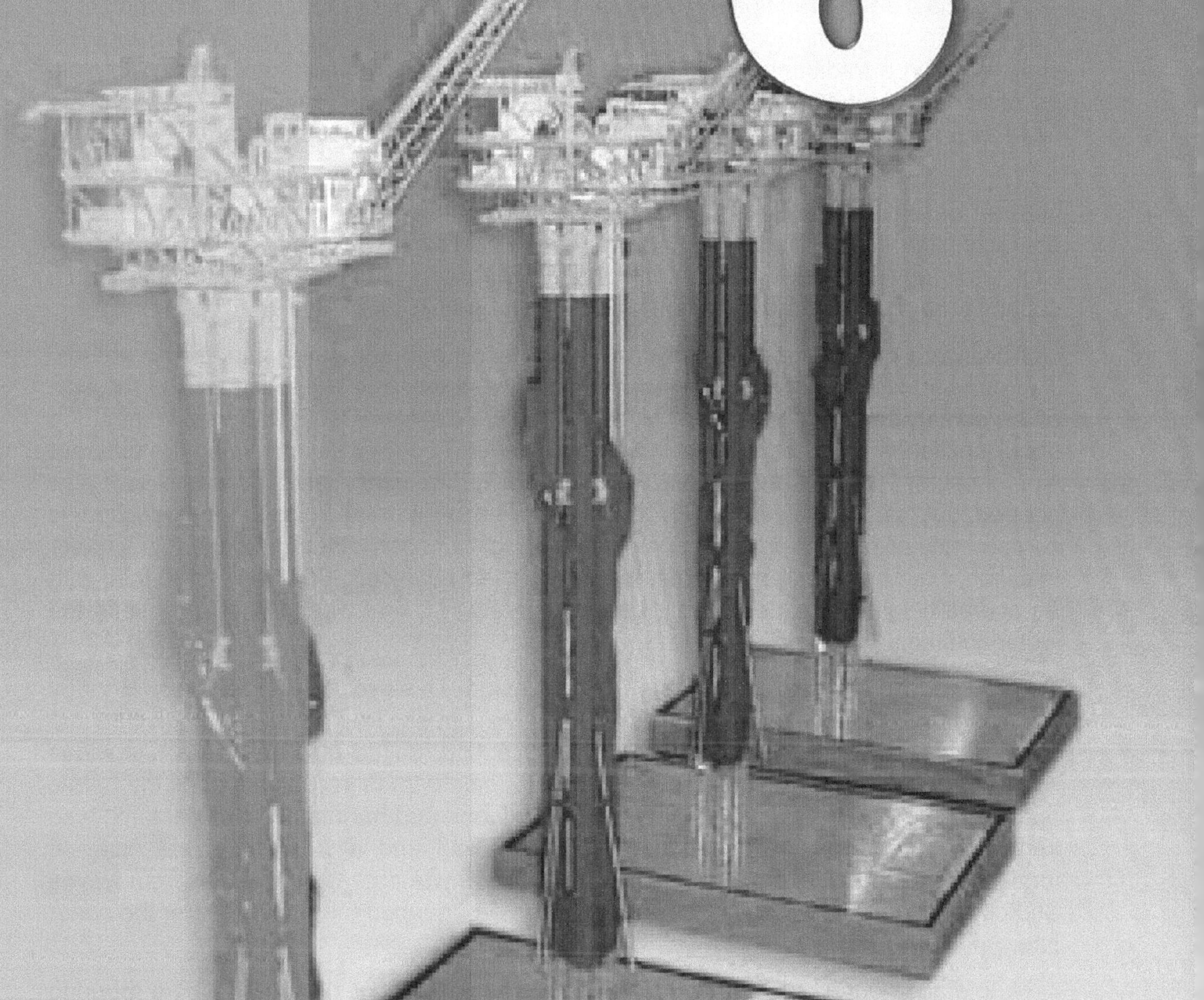

What

Is a Tension Spar Platform?

Tension Spar Platform은 무엇인가?

Lecture A What is a Tension Spar Platform?

A spar is a deep-draft floating caisson, which is a hollow cylindrical structure similar to a very large buoy. Its four major systems are hull, moorings, topsides, and risers. The spar relies on a traditional mooring system such as anchor-spread mooring to maintain its position. About 90 percent of the structure is underwater. Historically, spars were used as marker buoys, for gathering oceanographic data, and for oil storage.

The spar design is now being used for drilling, production, or both. The distinguishing feature of a spar is its deep-draft hull, which produces very favorable motion characteristics compared to other floating concepts. Low motions and a protected center well also provide an excellent configuration for deepwater operations. Water depth capability has been stated by industry as ranging up to 10,000 ft.

The hull is constructed by use of normal marine and shipyard fabrication methods. The number of wells, surface wellhead spacing, and facilities weight determine the size of the centerwell and the diameter of the hull. In the classic or full cylinder hull forms, the upper section is compartmentalized around a flooded centerwell containing the different type of risers. This section provides the buoyancy for the spar. The middle section is also flooded but can be economically configured for oil storage. The bottom section is compartmentalized to provide buoyancy during transport and to contain any field-installed, fixed ballast. Approximate hull diameter for a typical GOM spar is 130 feet, with an overall height, once deployed, of approximately 700 feet with 90% of the hull in the water column.

The first Spars were based on the Classic design. This evolved into the Truss Spar by replacing the lower section of the caisson hull with a truss. The Truss Spar is divided into three distinct sections. The cylindrical upper section, called the "hard tank," provides most of the in-place buoyancy for the Spar. The middle truss section supports the heave plates and provides separation between the keel tank and hard tank. The keel tank, also known as the "soft tank," contains the fixed ballast and acts as a natural hang-off location for export pipelines and flowlines since the environmental influences from waves and currents and associated responses are less pronounced there than nearer the water line.

A lateral catenary system of 6 to 20 lines keeps the spar on location. The mooring lines are a combination of spiral strand wire and chain. Because of its low motions, the spar can use a taut mooring system at a reduced scope and cost compared with a full catenary system. Each mooring line is anchored to the seafloor with a driven or suction pile.

hollow 비어있는 not solid; having a space or gap or cavity **buoy** 부이, 부표 a float attached by rope to the seabed to mark channels in a harbor or underwater hazards **configuration** 외형, 구성 an arrangement of parts or elements **capability** 능력, 소질 the quality of being capable **compartmentalize** 구분하다 separate into isolated compartments or categories **lateral** 측면의 **catenary system** 현수선시스템 **spiral** 나선의 in the shape of a coil

The hull end of the line passes through a fairlead located on the hull below the water surface, then extends up the outside of the hull to chain jacks at the top, usually 50 ft or more in elevation. Excess chain is stored in the hull. Depending on hull size and water depth, the moorings can vary in number up to 20 lines and contain 3,700 ft of chain and wire. Starting at the seafloor, a typical mooring leg may consist of approximately 200-ft long, 84-inch diameter piles; 200 ft of 4 inch bottom chain; 2,500 ft of 4 inch spiral strand wire; and 1,000 ft of 4 inch platform chain. The footprint created by the mooring system can reach a half-mile or more in diameter measured on centers from the hull to the anchor piles.

The topsides configurations follow typical fixed platform design practices. The decks can accommodate a full drilling rig (3,000 hp) or a workover rig (600-1,000 hp) plus full production equipment. Production capacities range up to 100,000 BOPD and 325 MMcfgpd. The type and scale of operation directly influence deck size. The larger topsides would be consistent with drilling, production, processing, and quarters facilities, and could also include remote wells/fields being tied back to the spar for processing. Total operating deck load, which includes facilities, contained fluids, deck structural and support steel, drilling/workover rig, and workover variable loads, can be 6,600 tons or more. Crew quarters on a production/workover spar may accommodate 18 workers, while a full drilling and production facility may house 100 people.

There are three basic types of risers: production, drilling, and export/import. Firstly, in production, each vertical access production riser is top tensioned with a buoyant cylinder assembly through which one or two strings of well casing are tied back and the well completed. This arrangement allows for surface trees and a surface BOP for workover. The drilling riser is also a top-tensioned casing with a surface drilling BOP, which allows a platform-type rig to be used. Lastly, export/import risers can be flexible or top-tensioned steel pipe or steel catenaries. Production risers consist of a conventional subsea wellhead at the mudline and a tieback connector with a stress joint for taking the stresses associated with environmentally imposed displacements.

elevation 고도, 해발 the highest level or degree attainable **configuration** 배치, 구성 **fairlead** 삭도기 **accommodate** 수용하다 provide with something desired or needed **BOP** blow out preventer - a large valve used in oil or natural gas drilling **BOPD** barrels of oil per day tied 묶여진 impose 부과하다 **displacement** 배수량 the weight of the water that a ship displaces when it is floating with its fuel tanks full and all stores aboard.

After you read

1. Up to what range of water depth capability does the industry state?
   (a) 1,000 ft
   (b) 10,000 ft
   (c) 1,000 m
   (d) 10,000 m

2. Please complete the sentence.
   (a) About 90 percent of the structure is underwater. Historically, spars were used as ____________, for gathering oceanographic data, and for oil storage
   (b) The spar design is now being used for _________, _________, or both. The distinguishing feature of a spar is its _________ hull, which produces very favorable motion characteristics compared to other floating concepts.
   (c) There are three basic types of risers: _________, ________, and ___________.
   (d) Excess chain is stored in the hull. Depending on hull size and water depth, the moorings can vary in number up to 20 lines and contain ________ ft of chain and wire.

3. Which one of the following is not true?
   (a) Approximate hull diameter for a typical GOM spar is 130 feet, with an overall height, once deployed, of approximately 700 feet (with 90% of the hull in the water column).
   (b) The cylindrical upper section, called the "hard tank," provides most of the in-place buoyancy for the Spar. A production riser conveys produced fluids from the well to the TLP surface production facilities
   (c) The keel tank, also known as the "hard tank," contains the fixed ballast and acts as a natural hang-off location for export pipelines and flowlines

Lecture c	Language Focus

1. Match the words with the definitions, then complete the definition with the given worlds.

Word	Definition
Drilling riser	a deep-draft floating caisson, which is a hollow ___________ structure similar to a very large buoy
Spar	can be flexible or top-tensioned steel pipe or steel __________.
Export/import riser	a _______________casing with a surface drilling BOP

catenaries   cylindrical   top-tensioned

2. Complete the sentences using the following words.

elevation   displacement   capability

1) Water depth _________ has been stated by industry as ranging up to 10,000 ft.
2) _____________ is the actual weight of the ship, since a floating body displaces its own weight in water.
3) It is located at an ____________ of 1,023 meters.

3. Complete the sentences using the following words.

1) The associated responses are less pronounced there ______nearer the water line.
   a) and   b) or   c) as   d) than
2) Approximate hull diameter for a typical GOM spar is 130 feet.
   a) In addition   b) Since   c) Why   d) Because

Lecture D Useful Expressions – both/neither/either

1. both / either/ neither

Both hotels are very good.
Neither hotel is expensive.
Either hotel is good for me. I don't care.

2. both of/ either of/ neither of

Both of the restaurants **are** good.
Neither of the restaurants **was** (or **were**) expensive.
Either of the books **is** available for borrowing in the library.

cf. Instead of "the" you can use those/these/my/John's
"both of the hotels" "both of those/these hotels" "both of John's hotels"

3. both and / either or / neither nor.

Both ~ and~	**Both** Tom **and** Bobby are late for class. I was **both** tired **and** exhausted when I finished my exam.
Neither ~ nor ~	**Neither** Susan nor **I** wasn't invited to the party. She said she would call me, but she **neither** wrote **nor** called.
either ~or ~	**Either** he **or** I have to work late tonight. **Either** you apologize, **or** I'll never speak to you again.

4. Put the proper words above.

a. Too much food is not good for your health. Too little food is not good for your health, either.
→ ________ too much food ______too little food is good for your health.
b. Mike doesn't like seeing a movie. He doesn't like watching TV, too.
→ Mike likes _____________seeing a movie ______________watching TV.
C. Billy bought a luxurious car last summer. Bob bought a luxurious car last summer, too.
→ __________ Blly _________ bought a luxurious car last year.

Lecture E Speak Out!!

1. Talk with your partner. What would you say in these situation? Complete the sentences using the expressions you've studied in this unit.

  Ilua hotel Lazza hotel	▪ Both _______ and ______ ____ nice hotels. ▪_________ The Ilua ________ The Lazza ______ not expensive compared with the others. ▪ I think _________ The Ilua ________ The Lazza ______ ok. I don't care.
	▪ Both a dog and a cat are animals. ▪_________ a dog ________ a cat can talk. ▪ You can have only one, _____ a dog ___ a cat.

2. Have a conversation with your partner based on the following example using Lecture D-B section and practice twice. You can use your own ideas.

Ex.

 Wow, Look at the pictures. They are really nice.

 Really? Thanks. Actually, those are being sent for "photo of the year contest'

 I see. Both the first and the second pictures are so funny.

 I haven't decided which one is going for the contest yet.

 Either the first one or the second one is ok.

 Ok, I am going for the second one.

Ex. candidate 1 candidate 2

Group name
- Big Top
-They are really popular!!

Dragon P Tiger G

Lecture F Episode on offshore!!

Work in pair. Ask and answer the question.

	What is a Spar?
	A spar is a deep-draft floating caisson, which is a hollow cylindrical structure similar to a very large buoy
	What are the distinguishing features of a Spar?
	The distinguishing feature of a spar is its deep-draft hull, which produces very favorable motion characteristics compared to other floating concepts. Low motions and a protected center well also provide an excellent configuration for deepwater operations. Water depth capability has been stated by industry as ranging up to 10,000 ft.
	What part can the Truss Spar be divided into?
	The Truss Spar can be divided into the cylindrical upper section, the middle truss section and the keel tank.
	Make your own!
	Make your own!

How the seas were named (2)

Dead Sea

Known as Mare Mortuum, or Dead Sea, by the Romans, this salty lake in Palestine supports practically no life. The limpid blue-green water extends 46 miles in length by 10 miles wide, falls to a depth of up to 1.300 feet and is 26 percent salt as opposed to the normal 3-4 percent in normal sea water.

Mediterranean Sea

The word comes from the Latin word medius, middle, and terra, earth. Mediterranean therefore seems to mean 'the sea in the middle of earth' which indeed the ancients believed it to be.

Pacific Ocean

The Portuguese navigator Ferdinand de Magalhes, better known to the world as Magellan, named this ocean on his voyage to the new world in 1520. Magellan meet gales and bad weather while sailing the Atlantic, and when he entered the Pacific he found a great body of water which was placid for several days during sailing.

He named it El Pacifico, the peaceful one, not knowing that it covered one third of the earth's surface and gave birth to terrible storms and monster waves unequaled in any other body of water.

Red Sea

The Red Sea was also named by Romans and came from the words Mare Rubrum, however they took it from an old Semitic name, the reason for which is not certain.

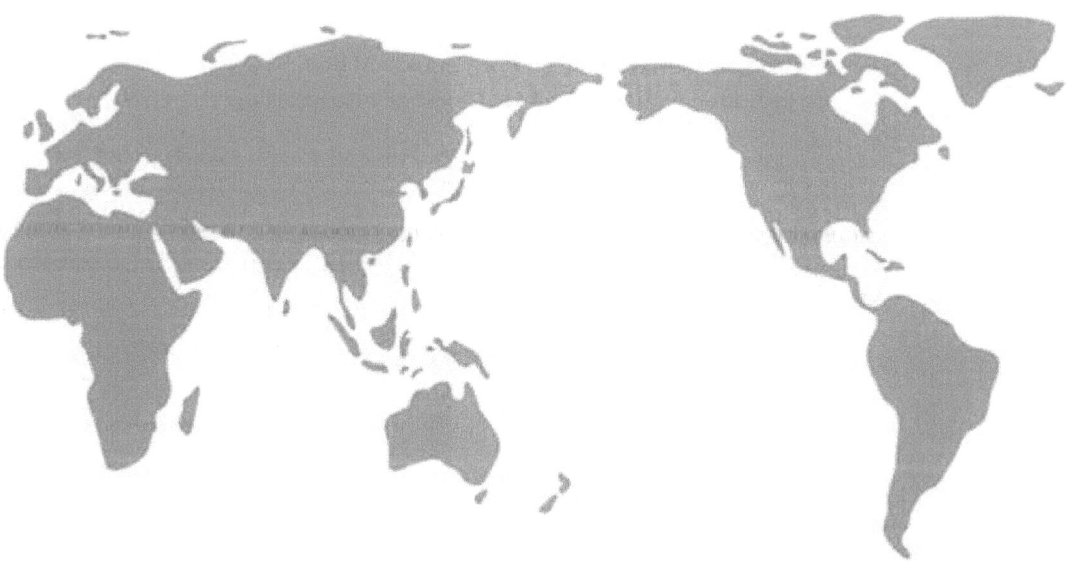

Pops English

Puff, the magic dragon

Puff, the magic dragon lived by the sea
And frolicked in the Autumn mist in a land called Honah-Lee
Little Jackie Paper loved that rascal Puff
And brought him strings and sealing wax and other fancy stuff

Chorus:
Puff, the magic dragon lived by the sea
And frolicked in the Autumn mist in a land called Honah-Lee
(repeat)

Together they would travel on a boat with billowed sail
Jackie kept a lookout perched on Puff's gigantic tail
Noble kings and princes would bow whene'er they came
Pirate ships would lower their flags when Puff called out his name, O

Chorus

A dragon lives forever but not so little girls and boys
Painted wings and giants' rings make way for other toys
One grey night it happened, Jackie Paper came no more
And Puff that mighty dragon, he ceased his fearless roar

His head was bent in sorrow, green scales fell like rain
Puff no longer went to play along the cherry lane
Without his life-long friend Puff could not be brave
So Puff that mighty dragon sadly slipped into his cave

Chorus

Puff the magic dragon LIVES by the sea (present tense!)
And frolicks in the Autumn mist in a land called Honah-Lee

The shortest answer is doing.

- Lord Herbert

UNIT 7

What Is a Drill Ship?

Drilling ship은 무엇인가?

Lecture A

What Is a Drill Ship?

A drillship is a maritime vessel that has been fitted with drilling apparatus. It is most often used for exploratory offshore drilling of new oil or gas wells in deep water or for scientific drilling. The drillship can also be used as a platform to carry out well maintenan ce or completion work such as casing and tubing installation or subsea tree installations. While drillships look similar to a tanker or cargo vessel, there are a couple of major differences. Drillships are equipped with a drilling derrick and moon pool. Additionally, drillships have extensive mooring or positioning equipment, as well as a helipad to receive supplies and transport staff. Typically employed in deep and ultra-deep waters, drillships work in water depths ranging from 2,000 to more than 10,000 feet (610 to 3,048 meters). Drilling equipment is passed through the vessel's moon pool and connected to the well equipment below via riser pipe, a somewhat flexible pipe that extends from the top of the subsea well to the bottom of the drillships. Drillships are differentiated from other offshore drilling units by their easy mobility. While semisubmersible rigs can also drill in deep waters, drillships are able to propel themselves from well to well and location to location, unlike semisubs, which usually rely on an outside transport vessel to transfer them from place to place.

While capable of drilling in deep and ultra-deep waters, a disadvantage to using a drillship is its susceptibility to being agitated by waves, wind and currents. This is especially troublesome when the vessel is actually drilling, because the drillship is connected to equipment thousands of feet under the sea. A proper mooring system on a drillship is integral to drilling successful wells. In shallower waters, drillships are moored to the seafloor with anywhere from 6 to 12 anchors. Once the water depth becomes too deep, drillships depend on dynamic positioning systems to keep the vessel in place while drilling. DPS relies on several thrusters located on the fore, aft and mid sections of the ship, which are activated by an onboard computer that constantly monitors winds and waves to adjust the thrusters to compensate for these changes. Sometimes, both positioning systems are used.

Drillships were first developed in the late 1940s by marine architects. Created to overcome water depth challenges offshore California in the Pacific Ocean, the first drillship was a surplus U.S. Navy patrol craft that was equipped with cantilevered drilling

differentiate 구별하다 **susceptibility** 민감함, **marine(naval) architect** 조선설계자 A naval a rchitect is an engineer who is responsible for the design, construction, and/or repair of ships, bo ats, other marine vessels, and offshore structures, **incorporated** 병합한 **troublesome** 곤란한, 성가신 difficult to deal with **subsequent** 뒤이은, 연이은 **constantly** 끊임없이 **adjust** 조절하다 alter or regulate so as to achieve accuracy or conform to a standard **activated** 활성화된 **compensate** 보상하다

equipment. Seeing some success, the next drill ship incorporated a moon pool and an onboard drilling derrick. After its success, companies soon ordered the construction of subsequent drillships. Transocean, Pride, Seadrill, Frontier Drilling and Noble are a few of the companies that own and operate drillships globally. Oil companies use sonic equipment to search for oil that is located deep beneath the ocean's floor. Once they decide exactly where they want to drill, they use a MODU, or mobile offshore drilling unit, to dig the well.

Drillships are a particular type of MODU that is composed of a ship with a drilling rig on its top deck. These ships can go right to the drill site. To remain stabilized, they use propellers and anchors to correct for drift while the drilling is taking place. The drill in the drill ship operates through a hole in the ship's hull. Unlike some other kinds of MODUs, drill ships can operate in deep water. Drillships and other types of MODUs use something called a riser to drill down into the ocean floor. This is how drilling fluids can pass from the ocean floor to the rig. Engineers then lower a drill string down through the riser. At the sea floor, a BOP (blowout preventer) clamps off the pipe in case of a blowout. Metal casings lined with cement walls are used to stabilize the well and keep it from collapsing. When a drill ship or any other kind of MODU hits oil, the well needs to be plugged up. Engineers put down one plug, which is held down by seawater or drilling mud until they can put a second one in place. After they cap the well, a permanent rig comes in to capture the oil, or in some cases the MODU is converted to a production rig and does the capturing itself.

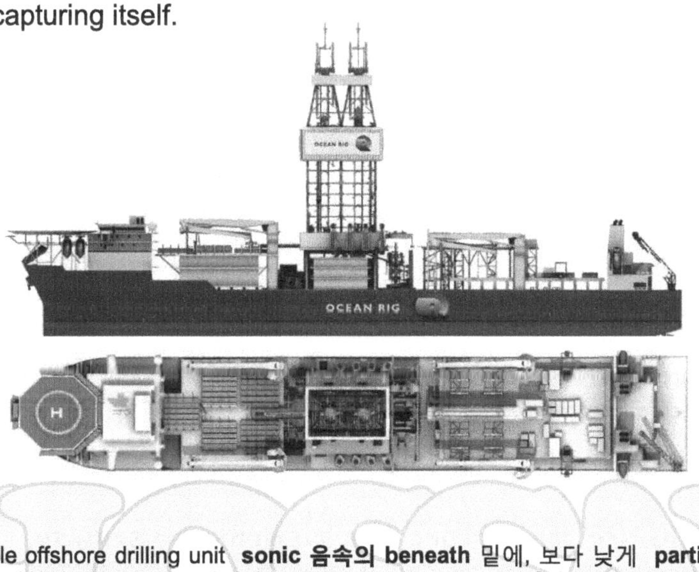

MODU mobile offshore drilling unit **sonic** 음속의 **beneath** 밑에, 보다 낮게 **particular** 특별히 **compose** 구성하다 **stabilize** 안정되다 **drilling fluid** a mixture of clays and chemicals and water, pumped down the drill pipe to lubricate and cool the drilling bit and to flush out the cuttings and to strengthen the sides of the hole

Lecture B	After you read

1. What was the first drillship?

   (a) A navy patrol craft that was equipped with cantilevered drilling equipment

   (b) A converted shuttle tanker with drilling equipment

   (c) A offshore construction vessel with drilling equipment

   (c) A diving support vessel

2. Typically employed in deep and ultra-deep waters, in what water depth do drillships work?

   (a) From 2,000 to more than 10,000 feet

   (b) From 200 to more than 1,000 feet

   (c) From 2,000 to more than 10,000 m

   (d) From 200 to more than 1,000 m

3. Please complete the sentence.

   (a) Drillships are equipped with a ______________and ____________.

   (b) Drilling equipment is ____________ the vessel's moon pool and connected to the well equipment below via _____________.

   (c) A disadvantage to using a drillship is its susceptibility to being agitated by ________, ________ and __________.

   (d) Once the water depth becomes too deep, drillships depend on ____________________to keep the vessel in place while drilling.

4. What is not the disadvantage of using a drillship?

   (a) Drillships are susceptible to being agitated by waves, wind and currents.

   (b) Drillships are more prone to movement in a seaway.

   (c) Drillships are subject to longer periods of downtime due to wind and wave action

   (d) Drillships can move from one location to the next rapidly.

Lecture c	Language Focus

1. Match the words with the definitions, then complete the definition with the given worlds.

Words	Definitions
Drill Fluid	an engineer who is _________for the design, construction, and/or repair of all kinds of ship
Drillship	pumped down the drill pipe to _______and cool the drilling bit and to flush out the cuttings and to __________the sides of the hole
naval architect	a maritime vessel that has been fitted with drilling _________

lubricate strengthen responsible apparatus

2. Complete the sentences using the following words.

troublesome activate differentiate

1) Drillships are _________ from other offshore drilling units by their easy mobility.
2) This is especially _________when the vessel is actually drilling.
3) They are ________ by an onboard computer that constantly monitors winds and waves.

3. Complete the sentences using the following words.

1) To remain _________ , they use propellers and anchors to correct for drift.
   a) stabilized   b) stabilizing   c) stabilizer
2) Drillships were first developed in the late 1940s ___ marine architects.
   a) through   b) to   c) in   d) By

Lecture D Useful Expressions – so/such

A. so and such

so + adjective/adverb	such + noun
so silly so beautiful **so** quickly so hard It is **so** nice outside . I like Ally and Bill. They are **so** nice.	**such** a story **such** people **such** a good day such a nice people It is **such** a bright day. I like Ally and Bill. They are **such** nice people.

B. so / such ~ that (*cf. **that*** is omissible.)

The book was **so** good **that** I couldn't put it down	It was **such a** good book **that** I couldn't put it down
I was **so** tired **that** I fell asleep in the sofa.	It was **such** nice weather **that** we spent the whole day on the beach.
I became **so** close to his family **that** it was hard to say good-bye.	He is **such a** mild man that no one had a quarrel with him.
Billy was injured **so** badly **that** he had to stay in hospital.	There was **such a** crowd that we could hardly move.

C. Put in so , such or such a..

1. It was a great vacation. We had _______ good time.
2. I couldn't believe the news. It was ______ shock.
3. The party was great. It was _____ shame you couldn't make it.
4. I like my next-door neighbors. They are _____ nice people.
5. She looked _______ tired all the time.
6. They've got _____much money that they don't know what to do with it.

Speak Out!!

1. Talk with your partner. What would you say in these situation? Complete the sentences using the expressions you've studied in this unit.

	so tired /couldn't move my feet at all
	such a lucky day/ can't sleep at all

2. Match each other and have a conversation with your partner based on the following example using Lecture D-B section and practice twice.

a. She worked hard. 	e. I didn't eat anything else for the rest of the day.	=	
b. I was tired. 	f. She made herself sick.	=	She worked so hard (that) she made herself sick.
c. It was terrible weather. 	g. I couldn't keep my eyes open.	=	
d. I had a big breakfast. 	h. We spent the whole day indoors.	=	

Lecture F **Episode on offshore!!**

Work in pair. Ask and answer the question.

	What is a Drill ship?
	A drillship is a marine vessel that's been modified to drill oil and gas wells.
	What is a difference between Drill ships and Semi-submersible rigs?
	Drillships are differentiated from other offshore drilling units by their easy mobility. While semisubmersible rigs can also drill in deep waters, drillships are able to propel themselves from well to well and location to location, unlike semisubs, which must rely on an outside transport vessel to transfer them from place to place.
	What was the first Drill ship?
	Answer the question!
	Make your own!
	Make your own!

The story of the Plimsoll Mark

In really old times of about 100-150 years ago, there were ships carrying so much cargo that they became known as 'coffin-ships'. Greedy ship owners wanted high profits and the more cargo put on board, the higher the profits would be.

They were then sent to sea and sometimes the ships overturned and the crew and ships were lost at sea. When a ship were lost, and it did frequently happen the ship owners profited handsomely as the ship were most of the time over insured.

The only people who lost, were the helpless crew of the fateful ships. Then, in 1824 a man named Samuel Plimsoll was born. He became a brewery manager in London and later turned in to become a coal dealer. He further made the British seaman's cause his own. Shortly after being elected to Parliament from Derby in 1968, Plimsoll tried passing a bill improving the horrific situation with seamen continuously being lost at sea.

But he failed his attempt. Instead he wrote a book named 'Our Seamen', which he published in 1872. It was a collection of dramatic stories he had been told by seamen down at the docks where the ships came in.

He was then sued for liability by certain ship owners and had to apologize in Parliament for revealing privileged information. However, as fate would have it, the book made an impact and soon the British people supported his views on greedy ship owners and their recklessness. The Parliament had to gather a commission to investigate the murderous claims put forward in the book.

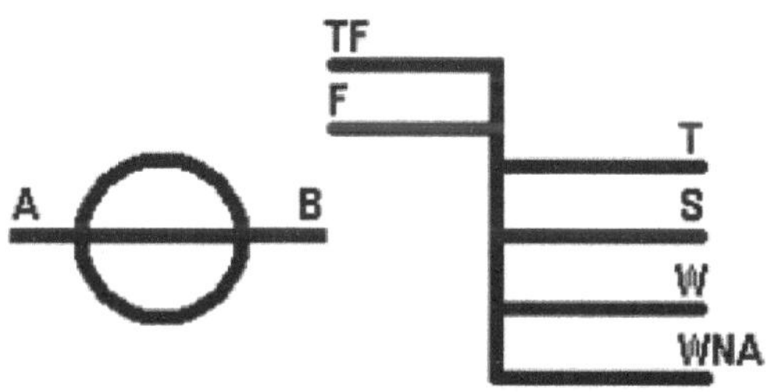

Pops English

Just the way you are

Oh her eyes her eyes
make the stars look like they're not shining
her hair her hair
falls perfectly without her trying.
she's so beautiful and I tell her every day

yeah I know, I know
when I compliment her she wont believe me
and its so, its so
sad to think she don't see what I see
but every time she asks me do I look okay , I say

When I see your face
there's not a thing that I would change
cause you're amazing just the way you are
and when you smile,
the whole world stops and stares for awhile
cause girl you're amazing
just the way you are

her lips, her lips
I could kiss them all day if she'd let me
her laugh, her laugh
she hates but I think it so sexy
she's so beautiful
and I tell her every day

oh you know, you know, you know
I'd never ask you to change
if perfect is what you're searching for
then just stay the same
so don't even bother asking
if you look okay you know I say

when I see your face
there's not a thing that I would change
cause you're amazing just the way you are
and when you smile,
the whole world stops and stares for a while
cause girl you're amazing
just the way you are

A likely impossibility is always preferable to an unconvincing possibility.

- Aristotle

UNIT 8

What
Is a Shuttle Tanker?

셔틀탱커는 무엇인가?

Lecture A

What is a Shuttle Tanker?

A shuttle tanker is a specialized ship designed to transport oil from offshore oil fields to onshore refineries. Shuttle tankers are often used as an alternative to pipelines in harsh climates, remote locations or deepwater. Shuttle tankers were initially used in the North Sea in the 1970s. Since then, the value of using a shuttle tanker instead of pipelines has spread worldwide.

Operators have recognized the flexibility of loading oil and transporting it to any destination they want as opposed to pipelines that are fixed to one receiving terminal. Operators also favor the ability of shuttle tankers to segregate the oil. Pipelines often blend crude from various fields. Shuttle tankers are also easier to maintain and have a back-up-system in place if they require maintenance. If a pipeline requires maintenance, the whole network is likely to be shut in. Shuttle tankers have also proven to be safer than conventional crude oil tankers. Conventional tankers can only be used in calm waters. They load from offshore installations with assistance from tug boats to stay on location; whereas shuttle tankers operate independently in all water and weather conditions. Shuttle tankers are equipped with two or three bow thrusters and stern thrusters as well as dynamic positioning to keep the tanker on location. Flap rudders and controllable pitch propellers also help to increase low speed maneuverability.

Field operators often award shuttle tanker owners long-term contracts to offload and transport oil from the field installation to the receiving terminal. The crew is highly trained to maneuver the tanker safely when close to offshore installations. They also go through extensive training to operate the offshore loading system and the dynamic positioning system safely. The offshore loading system transfers oil from a FPSO. The loading system is on the bow of the ship, which allows the shuttle tanker to load safely from various platforms, regardless of weather conditions. A special cone-shaped loading system can also be installed on the keel of the ship. The offshore platform has a field export system that connects to the loading system on the shuttle tanker. The type of export system onboard varies based on safety, weather conditions, water depth, seabed conditions and cost.

refinery 정유소 an industrial plant for purifying a crude substance **segregate** 분리하다 separate or isolate (one thing) from another **blend** 섞다 combine into one **tug boat** 예인선 a powerful small boat designed to pull or push larger ships **maneuverability** the quality of being maneuverable **bow/stern thruster 선수/선미추진기** a engine that provides the thrust needed to maneuver a ship **keel** 용골 one of the main longitudinal beams of the hull of a vessel

The single point mooring system (SPM) is an offshore loading buoy where tankers can connect to load and offload oil. SPMs can handle any size ship. The body of the buoy, mooring and anchoring elements, product transfer system and other components make up an SPM. Together, these parts allow shuttle tanker operators to favorably position the vessel around the mooring point so it is less affected by weather conditions. A hose from the export system is then connected to the loading system on the vessel. Dynamic positioning systems have significantly reduced wear-and-tear on SPM equipment. After the introduction of dynamic positioning in the 1980s, the Ugland Kongsberg Offshore Loading System was introduced. This less expensive option can only be used with dynamically positioned shuttle tankers.

Eight years later, the Submerged Turret Loading (STL) system hit the market in 1993. An STL has a submerged buoy that comes through a cone-shaped turret on the keel of the shuttle tanker. The STL is used in the most extreme weather conditions with sea states up to 6 meters. Introduced in the late 1990s, Single Anchor Loading (SAL) is used primarily in shallow water and in good weather conditions. The SAL is a simple, cost effective alternative to the STL. The Tandem Loading System(TLS), first carried out in 1991, requires the shuttle tanker to connect to the stern of the FPSO or FSO. Dynamic positioning is necessary to monitor the positions of two units.

handle 다루다 **component** 구성요소 an abstract part of something **affect** 영향을 미치다 have an effect upon **reduce** 줄이다 **wear-and-tear** 손상, 마모 decrease in value of an asset due to obsolescence or use **element** tandem 앞뒤로 연결된 **significantly** 매우 의미있게, 현저하게 **extreme** 극도의 of the greatest possible degree or extent or intensity **primarily** 본래,주로 mainly **effective** 효과적인 **alternative** 대체의, 대안 **carry out** 실행/실시하다

Lecture B

After you read

1. What kind of system on the shuttle tanker is contributing to significantly reduce wear-and-tear on SPM equipment?
   (a) Dynamic positioning system
   (b) Spread mooring system
   (c) Anchoring system
   (d) Bow loading system

2. What does SPM stand for?
   (a) Single Point Mooring
   (b) Single Positioning Monitoring
   (c) Speed Power Monitoring
   (d) Screw Propeller Monitoring

3. Please complete the sentence.
   (a) A shuttle tanker is a specialized ship designed to transport oil from ________ oil fields to ________ refineries.
   (b) Shuttle tankers are equipped with two or three bow ________ and stern ________ as well as dynamic positioning to keep the tanker on location.
   (c) Flap rudders and ____________________propellers also help to increase low speed maneuverability.

4. What is not the main characteristics of shuttle tankers?
   (a) A shuttle tanker is a specialized ship designed to transport oil from offshore oil fields to onshore refineries.
   (b) Shuttle tankers are often used as an alternative to pipelines in harsh climates, remote locations or deepwater.
   (c) The Tandem Loading System requires the shuttle tanker to connect to the stern of the FPSO or FSO.
   (d) Shuttle tankers load from offshore installations with assistance from tug boats to stay on location.

Language Focus

1. Match the words with the definitions, then complete the definition with the given worlds.

STL	offshore ______________ where tankers can connect to load and offload oil.
SPM	requires the shuttle tanker to connect to the __________ of the FPSO or FSO
TLS	submerged buoy that comes through a _______________ on the keel of the shuttle tanker.

loading buoy　　stern　　cone-shaped turret

2. Complete the sentences using the following words.

maneuver　　bow　　regardless of　　highly

1) The crew is ___________ trained to ____________ the tanker safely when close to offshore installations.
2) The loading system is on the _____ of the ship, , which allows the shuttle tanker to load safely from various platforms, _________________ weather conditions.

3. Complete the sentences using the following words.

1) The Submerged Turret Loading (STL) system ______ the market in 1993.
   a) crashed　　b) hit　　c) punched
2) Shuttle tankers ___________to be safer than conventional crude oil tankers.
   a) had proven　　b) was proven　　c) have proven

Lecture D Useful Expressions – It is said that ~

1. It is said that ~/ He is said to be ~

It is said that he is 105 years old. = He is said to be 105 years old.	=	People say that he is 105 years old.

alleged believed reported known expected considered thought

■ The police are looking for a missing girl.

It is believed that she is wearing a white dress and red shoes. — She is believed to be wearing a white dress and red shoes.

■ The strike started two weeks ago.

It is expected that it will end soon. — The strike is expected to end soon.

■ A friend of mine has been arrested.

It is alleged that he stole a car. — He is alleged to have stolen a car.

■ The police are investigating that area.

It is reported that two people were injured in the explosion. — Two people are reported to have been injured in the explosion.

2. Make sentences in two ways above.

a. They reported that the man got lost in the mountain.

= It was ______________________________ .

= The man ______________________________ .

b. We know that a friend in need is a friend indeed.

=

=

c. They considered that some of the employees were against the new policy.

=

=

Lecture E	Speak Out!!

A. Talk with your partner. What would you say in these situation? Complete the sentences using the expressions you've studied in this unit.

	▪ Hanna likes drinking so much. She often gets easily drunken. Her parents worry about her all the time. They believe that she was drunken last night again. = =
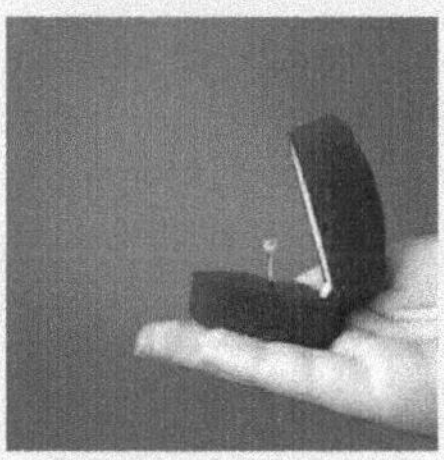	▪ Anna and her boyfriend have been together for 3 years. She expects that he will propose to her soon. = =

B. Have a conversation with your partner based on the following example using Lecture D and practice twice. You can use your own ideas.

There are a lot of rumors about him. Here are what people say.

EX. 1. It is said that he is a millionaire.
= He is said to be a millionaire.
Practice more using *believe, think, consider, etc* whatever you like it.

Lecture F **Episode on offshore!!**

Work in pair. Ask and answer the question.

	What is a shuttle tanker?
	A shuttle tanker is a specialized ship designed to transport oil from offshore oil fields to onshore refineries.
	Why do the operators favor the shuttle tanker?
	Operators favor the ability of shuttle tankers to segregate the oil. Pipelines often blend crude from various fields. Shuttle tankers are also easier to maintain and have a back-up-system in place if they require maintenance. If a pipeline requires maintenance, the whole network is likely to be shut in.
	What is a SPM?
	Answer the question!
	*Make your own!*
	*Make your own!*

OUTRO

Viking tales, true or false

When the Norwegian Vikings first came to 'Hope', the land which is now known as New York, they brought gifts to the locals. They gave them milk, but the problem is that the locals had never tasted it before. Due to it's foul condition and taste, the locals thought the Vikings were trying to poison them and chased them back to Newfoundland. This is the reason why Viking never settled in Hope (New York).

When Christopher Columbus landed in the new world in 1492, he reported back to Queen Isabella that they all spoke Norwegian (this is probably not true):

Dear Queen Isabella, I have two things to tell you, one good news and one bad news. The good news is that we did fine the New World. The bad news is that all the people there already speak Norwegian.

When the Oseberg Viking ships, now on display at the Viking ship museum in Oslo (Norway), were put to the test in powerful and high tech -computers there was little or nothing that could be done to improve their construction and seaworthiness.

Pops English

Rhythm of Falling Rain

Listen to the rhythm of the falling rain
Telling me just what a fool I've been
I wish that it would go and let me cry in vain
And let me be alone again

Now the only girl I care about has gone away
Looking for a brand new start
But little does she know that when she left that day
Along with her she took my heart

Rain please tell me know does that seem fair
For her to steal my heart away when she don't care
I can't love another
When my heart's somewhere far away

The only girl I care about has gone away
Looking for a brand new start
But little does she know that when she left that day
Along with her she took my heart

Rain won't you tell her that I love her so
Please ask the sun to set her heart aglow
Rain in her heart
And let the love we knew start to grow

Listen to the rhythm of the falling rain
Telling me just what a fool I've been
I wish that it would go and let me cry in vain
And let me be alone again

Oh listen to the falling rain
Pitter-patter pitter-patter

Listen listen to the falling rain
Pitter-patter pitter-patter

If you knew how much work went into it, you wouldn't call it genius.

- Michelangelo

UNIT 9

What Is a FLNG?

FLNG는 무엇인가?

Lecture A

What is FLNG?

Generically, there are three types of "floating LNG" currently implemented or being planned.

Floating Regasification: As used on floating storage and regas units (FSRUs) and shuttle regas vessels (SRVs). FSRUs are designed to receive LNG from a shuttle conventional LNG vessel, store and then re-gasify the LNG as and when required. The FSRUs may or may not also be designed to operate as a conventional LNG vessel, and may sail to pick up and deliver LNG cargoes anywhere in the world , within operating and insurance limits. FSRUs that are barge -like are unlikely to perform such ocean voyages. SRVs are designed to perform differently - they act more like a conventional LNG vessel, picking up and delivering cargoes, except that the regas delivery function is typically performed at a buoy located some distance from the shore. The intention being that the SRV will re-gasify and unload its cargo as soon as possible upon arrival at the remote buoy. The gas is then transported (via undersea pipeline) to gas storage facilities on-shore. Accordingly, SRVs are not intended to provide the LNG storage function of an FSRU. In both FSRUs and SRVs there is no gas liquefaction process.

On-board Reliquefaction: As a means of cargo preservation, on -board reliquefaction equipment has been installed on conventional LNG vessels. The reliquefaction equipment reliquefies LNG boil –off gas, with the resultant LNG being pumped back into the vessel's containment system. While this system requires additional power input, it can be a useful means of preserving the value of the vessel's LNG cargo.

Floating Liquefaction: Involves the production of LNG directly at offshore oil and gas fields by floating production storage and offloading (FPSO) type vessels. Of the three types of "floating LNG", it is fair to say that floating liquefaction presents the most difficult technical, commercial, legal and operating issues.

While the development and implementation of FLNG technology seems within grasp, the technology has not yet been successfully implemented on a commercial scale. Challenges include: the integration of the topside processing equipment with the hull which is with a deck layout that meets all operational and safety requirements, selection of a containment system which has the difficulty of constantly changing volumes within the tanks and related sloshing issues, "maximization" of equipment, vessel mooring

Regasification 재기화 **liquefaction** 액화 the conversion of a solid or a gas into a liquid **intention** 의도 an act of intending; a volition that you intend to carry out **grasp** 꽉 쥐다 hold firmly **containment** 억제 봉쇄 the act of containing; keeping something from spreading **reliquefaction** 재액화 **implementation** 이행 the act of accomplishing some aim or executing some order system up-scaling of plant capacity and ship -to-ship offloading of LNG, particularly in dynamic metocean conditions.

Consider the logistical issues relevant to operating a "floating port," potentially located far from landbased services, without drydocking, for a period of 20 years or more. Relevant issues may include: the provision of tug services if required, the scheduling of FLNG vessel and equipment maintenance , the delivery and storage of spare parts, fuels and catalysts, the scheduling of gas receipt, processing and liquefaction to meet LNG and NGL delivery obligations and the scheduling of liquid cargo transfers (both LNG and NGLs), in each case, in potentially dynamic metocean conditions. Recognizing that failure at any stage of the process could cause contracted cargo delivery obligations to be breached, a shut -down of the FLNG vessel and, potentially, a "shut -in" of the field.

Figure: Potential FLNG value chain

source from : DSME

metocean an abbreviation of the two words "Meteorology" and "Oceanography". offshore industry to describe the physical environment say near an offshore platform **provision** 제공 a store or supply of something **logistical=logistic** 물류의 **potentially** 잠재적으로 **relevant** 적절한 **obligation** 의무 the social force that binds you to the courses of action demanded by that force;

Lecture B

After you read

1. What is not the type of "floating LNG" currently implemented or being planned?
   (a) Floating Regasfication
   (c) On-board Reliquefaction
   (c) Floating Liquefaction
   (d) On-bard Liquefaction

2. Which type of floating LNG is designed to receive LNG from a shuttle conventional LNG vessel, store and then regasify the LNG as and when required?
   (a) FLNG
   (b) FSRU
   (c) FPSO
   (d) LNG RV

3. Which type of floating LNG involve the production of LNG directly at offshore oil and gas fields by floating production storage and offloading ?
   (a) FLNG
   (b) FSRU
   (c) FPSO
   (d) LNG RV

4. Please complete the sentence.
   (a) The reliquefaction equipment reliquefies LNG ________gas, with the resultant LNG being pumped back into the vessel' s containment system.
   (b) SRVs are not intended to provide the LNG ________ function of an FSRU.
   (c) In both FSRUs and SRVs there is no gas ________________process.

Language Focus

1. Match the words with the definitions, then complete the definition with the given worlds.

Words	Definitions
FSRUs	an abbreviation of the two words ________________ and ________________
Floating Liquefaction	the production of LNG ______at offshore oil and gas fields by ____________ vessels
metocean	receive LNG from a shuttle conventional LNG vessel, store and then ____________the LNG

oceanography   directly   re-gasify   meteorology   FPSO type

2. Complete the sentences using the following words.

failure   potentially   logistical   obligations

1) Consider the ________issues relevant to operating a "floating port," ______________ located far from land-based services.

2) Recognizing __________that at any stage of the process could cause contracted cargo delivery __________ to be breached.

3. Complete the sentences using the following words.

1) ______ this system requires additional power input, it can be a useful means of preserving the value of the LNG cargo.

a) Also   b) As   c) While   d) Because

2) ___ the three types of "floating LNG" , it is fair to say that floating liquefaction presents the most difficult issues.

a) Of   b) Since   c) Out   d) During

Lecture D Useful Expressions – speak to you/ mention it

A. Verb + preposition + Objective/ Verb + Objective

Verb + preposition +O		
speak /talk	to	He will **speak to** you.
listen	to	**Listen to** me, son.
reply	to	Please, **reply to** my letter.
react	to	How did she **react to** my suggestion? She **reacted** negatively **to** my suggestion.
talk about	about	Can you **talk about** your problem?
account	for	His careless driving **accounted for** the accident
agree	with	I **agree with** you.
apologize	to s.b for s.t	I **apologize** to her **for** my rude action.

Verb +O	
mention	Don't **mention** it.
discuss	I will **discuss** this issue later.
instruct	She **instructed** me that her father was upstairs.
explain	Let me **explain** this.
answer	He didn't **answer** the question.
approve	We cannot **approve** his proposal.
oppose	We **opposed** violence to violence
call/phone/telephone	Please, **call** me. / **Phone** me the news.
ask	He **asked** a policeman the way.

A. Talk with your partner. What would you say in these situation? Complete the sentences using the expressions you've studied in this unit.

2. Talk with your partner. What would you say in these situation? Complete the sentences using the expressions you've studied in this unit.

Ex.

Dad, I need to talk to you.	Ok. I am listening.
I am 16 , so I want a car.	What are you talking about?
Let me explain this first.	No. We'll never discuss this!!

Those are son's wishes.

Lecture F

Episode on offshore!!

Work in pair. Ask and answer the question.

	What is the FSRU?
	FSRUs are designed to receive LNG from a shuttle conventional LNG vessel, store and then re-gasify the LNG as and when required.
	What can be inferred about floating liquefaction?
	Answer the question!
	What kind of issues does the author consider in related to logistics?
	Answer the question!
	*Make your own!*
	*Make your own!*

MIDGARDSORMEN

VIKING LAWS

§1 BE BRAVE AND AGGRESSIVE
BE DIRECT
GRAB ALL OPPORTUNITIES
USE VARYING METHODS OF ATTACK
BE VERSATILE AND AGILE
ATTACK ONE TARGET AT A TIME
DON'T PLAN EVERYTHING IN DETAIL
USE TOP QUALITY WEAPONS

§2 BE PREPARED
KEEP WEAPONS IN GOOD CONDITIONS
KEEP IN SHAPE
FIND GOOD BATTLE COMRADES
AGREE ON IMPORTANT POINTS
CHOOSE ONE CHIEF

§3 BE A GOOD MERCHANT
FIND OUT WHAT THE MARKET NEEDS
DON T PROMISE WHAT YOU CAN T KEEP
DON T DEMAND OVERPAYMENT
ARRANGE THINGS SO THAT YOU CAN RETURN

§4 KEEP THE CAMP IN ORDER
KEEP THINGS TIDY AND ORGANIZED
ARRANGE ENJOYABLE ACTIVITIES WHICH STRENG-
THEN THE GROUP
MAKE SURE EVERYBODY DOES USEFUL WORK
CONSULT ALL MEMBERS OF THE GROUP
FOR ADVICE

Pops English

Sorry Seems to be the hardest word

What have I got to do to make you love me
What have I got to do to make you care
What do I do when lightning strikes me
And I wake to find that you're not there

What do I do to make you want me
What have I got to do to be heard
What do I say when it's all over
And sorry seems to be the hardest word

It's sad, so sad
It's a sad, sad situation
And it's getting more and more absurd
It's sad, so sad
Why can't we talk it over
Oh it seems to me
That sorry seems to be the hardest word

It's sad, so sad
It's a sad, sad situation
And it's getting more and more absurd
It's sad, so sad
Why can't we talk it over
Oh it seems to me
That sorry seems to be the hardest word

What do I do to make you love me
What have I got to do to be heard
What do I do when lightning strikes me
What have I got to do
What have I got to do
When sorry seems to be the hardest word

Many of life's failures are people who did not realize how close they were to success when they gave up

- Thomas A. Edison -

UNIT 10

How does Offshore Pipeline installation work?

어떻게 Offshore 파이프를 설치하는가?

Lecture A How is Pipeline Laying?

Laying pipe on the seafloor can cause a number of challenges, especially if the water is deep. There are three main ways that subsea pipe is laid -- S-lay, J-lay and tow-in -- and the pipelay vessel is integral to the success of the installation.

Buoyancy affects the pipelay process, both in positive and negative ways. In the water, the pipe weighs less if it is filled with air, which puts less stress on the pipelay barge. But once in place on the sea bed, the pipe requires a downward force to remain in place. This can be provided by the weight of the oil passing through the pipeline, but gas does not weigh enough to keep the pipe from drifting across the seafloor. In shallow-water scenarios, concrete is poured over the pipe to keep it in place, while in deepwater situations, the amount of insulation and the thickness required to ward of hydrostatic pressure is usually enough to keep the line in place. While jumpers are typically short enough to be installed in sections by ROVs, flowlines and pipelines are usually long enough to require a different type of installation, whether that is tow-in, S-lay or J-lay.

Tow-In Pipeline Installation

Tow-in installation is just what it sounds like; here, the pipe is suspended in the water via buoyancy modules, and one or two tug boats tow the pipe into place. Once on location, the buoyancy modules are removed or flooded with water, and the pipe floats to the seafloor.

There are four main forms of tow-in pipeline installation. The first, the surface tow involves towing the pipeline on top of the water. In this method, a tug tows the pipe on top of the water, and buoyancy modules help to keep it on the water's surface. Secondly, using less buoyancy modules than the surface tow, the mid-depth tow uses the forward speed of the tug boat to keep the pipeline at a submerged level. Once the forward motion has stopped, the pipeline settles to the seafloor. Thirdly, off-bottom tow uses buoyancy modules and chains for added weight, working against each other to keep the pipe just above the sea bed. When on location, the buoyancy modules are removed, and the pipe settles to the seafloor. Lastly, the bottom tow drags the pipe

lay 놓다 **integral** 완전한, 필수적인 existing as an essential constituent or characteristic **pour** 붓다 **downward** 아래로 향하여 **hydrostatic** 유체 정역학적인 relating to fluids at rest or to the pressures they exert or transmit **weigh** 무게를 재다 **insulation 절연제, 단열제** **flowline** used on a drilling rig, is a large diameter pipe that is connected to the bell nipple(under the drill floor) and extends to the possum belly (on the mud tanks) and acts as a return line, (for the drilling fluid as it comes out of the hole), to the mud tanks.

along the sea bed, using no buoyancy modules. Only performed in shallow-water installations, the sea floor must be soft and flat for this type of installation.

S-Lay Pipeline Installation

When performing S-lay pipeline installation, pipe is eased off the stern of the vessel as the boat moves forward. The pipe curves downward from the stern through the water until it reaches the "touchdown point," or its final destination on the seafloor. As more pipe is welded in the line and eased off the boat, the pipe forms the shape of an "S" in the water.

J-Lay Pipeline Installation

Overcoming some of the obstacles of S-lay installation, J-lay pipeline installation puts less stress on the pipeline by inserting the pipeline in an almost vertical position. Here, pipe is lifted via a tall tower on the boat, and inserted into the sea. Unlike the double curvature obtained in S-lay, the pipe only curves once in J-lay installation, taking on the shape of a "J" under the water.

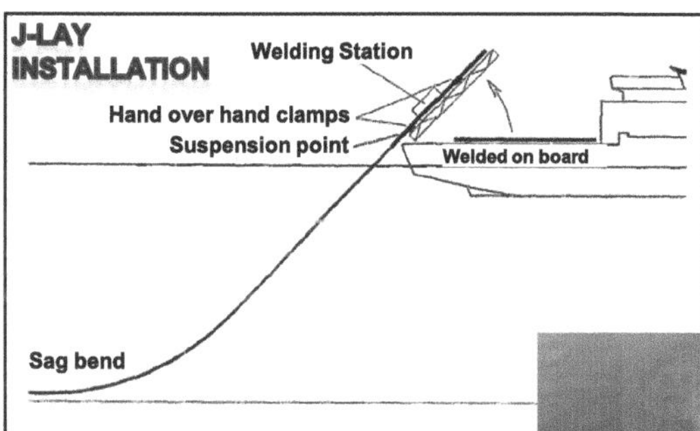

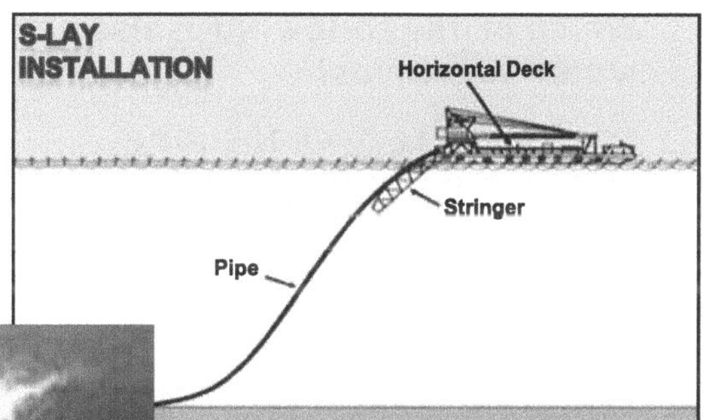

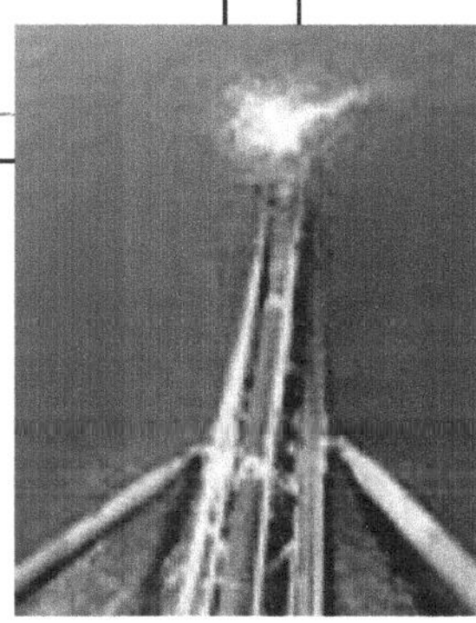

flat 편편한 having a surface without slope **curve** 곡선 **destination** 목적지 **ease off** 천천히 움직이다 **overcome** 극복하다, 정복하다 deal with successfully **obstacle** 장애물 an obstruction that stands in the way **vertical** 수직의 **curvature** 곡률, 만곡 **stress** 응력 force that produces strain on a physical body

Lecture B **After you read**

1. Which is not one of the three main ways that subsea pipe is laid?
   (a) S-lay
   (b) J-lay
   (c) Tow-in
   (d) T-lay

2. What word can substitute for the word "integral" in the second sentence of the first paragraph?
   (a) dangerous
   (b) essential
   (c) fundamental
   (d) early

3. Which of the following is used to keep the pipe in place in shallow-water scenarios?
   (a) concrete over the pipe
   (b) insulation
   (c) hydrostatic pressure
   (d) Steel tubular

4. Which is not true about the article?
   (a) Tow-in installation is that the pipe is suspended in the water via buoyancy modules, and one or two tug boats tow the pipe into place.
   (b) Pipe is eased off the stern of the vessel as the boat moves forward in S-lay pipeline installation.
   (c) J-lay pipeline installation puts less stress on the pipeline by inserting the pipeline in an almost horizontal position.
   (d) In the water, the pipe weighs less if it is filled with air, which puts less stress on the pipelay barge.

Lecture c	Language Focus

1. Match the words with the definitions. then complete the definition with the given worlds.

Word		Definition
flowline	●	relating to ________ at rest or to the pressures they exert or _____
hydrostatic	●	________that produces ________ on a physical body
stress	●	a large _____________pipe that is connected to the bell nipple

transmit   fluids   strain   diameter   force

2. Complete the sentences using the following words.

positive   weigh   ease off   negative

1) Buoyancy affects the pipelay process, both in _______and _______________ways.
2) The gas does not __________ enough to keep the pipe from drifting across the seafloor.
3) The pipe is ________________the stern of the vessel as the boat moves forward.

3. Complete the sentences using the following words.

1) Only ___________in shallow-water installations, the sea floor must be soft and flat for this type of installation.
   a) performing   b) performed   c) had performed
2) Tow-in installation is just ________ it sounds like.
   a) how b) what   c) When d) which

Lecture D Useful Expressions - ~ing/P.P Phrases

1. ~ing/P.P Phrases

-When two things happen at the same time, you can use ~ing/P.P phrases.

Kate is in the kitchen and she is making coffee.

=Kate is in the kitchen **making** coffee.

-When one action is caused by another action, you can use ~ing/p.p phrases.

Because I was feeling tired, I went to bed early.

=**Feeling** tired, I went to bed early.

-When one action is a condition for another , you can use ~ing/p.p phrases.

If it is marketed cleverly, the product will be very popular.

=**Marketed** cleverly, the product will be very popular.

2 having + P.P Phrases

When one action happens after another , you can use having +P.P phrases.

After she finished her work, I went to bed early.

=**Having finished** her work, I went to bed early.

3. Make sentences using ~ing/P.P/Having P.P phrases

a. Because she doesn't have a car, she has trouble getting around.

b. They had dinner, and then they continued their trip.

c. I was walking home in the rain. I got very wet.

d. After she was offended by his remarks, she walked our of the room.

Lecture E	**Speak Out!!**

1. Talk with your partner. Match the proper sentences each other using the expressions you've studied in this unit.

EX. 1. Feeling tired, he lay down on the grass.

~~a. He felt tired.~~	He was getting off a bus.
b. Bobby slipped and fell.	~~He lay down on the grass.~~
c. Sue went to London.	They were hiking in the woods.
d. Two kids got lost.	They were drinking hot coffee.
e. All of the people watched the sunrise.	She looked at the sights and took pictures.

2. Have a conversation with your partner based on the following example using Lecture D and practice twice. You can use your own ideas.

These are what Kate did yesterday in order. She is a famous actress.

EX. a. I woke up at 7 am. b. Then, I went to the gym.
= Having woken up 7 am, I went to the gym.
Practice more using having ~ing.

Lecture F

Episode on offshore!!

Work in pair. Ask and answer the question.

	What is Tow -in pipeline installation?
	Tow-in installation is just what it sounds like; here, the pipe is suspended in the water via buoyancy modules, and one or two tug boats tow the pipe into place. Once on location, the buoyancy modules are removed or flooded with water, and the pipe floats to the seafloor.
	How is S-Lay pipeline installed?
	Answer the question!
	How is J-Lay pipeline installed?
	Answer the question!
	*Make your own!*
	*Make your own!*

Seven Sisters

The "Seven Sisters" was a term coined in the 1950s by businessman Enrico Mattei then-head of the Italian state oil company ENI, to describe the seven oil companies which formed the "Consortium for Iran" cartel and dominated the global petroleum industry from the mid-1940s to the 1970s. The group comprised Anglo-Persian Oil Company (now BP); Gulf Oil, Standard Oil of California (Socal) and Texaco (now Chevron); Royal Dutch Shell; and Standard Oil of New Jersey (Esso) and Standard Oil Company of New York (Socony) (now ExxonMobil). Prior to the oil crisis of 1973, the members of the Seven Sisters controlled around 85% of the world's petroleum reserves, but in recent decades the dominance of the companies and their successors has declined as a result of the increasing influence of the OPEC cartel and state-owned oil companies in emerging-market economies

In 1951 Iran nationalised its oil industry, then controlled by the Anglo-Iranian Oil Company (now BP), and Iranian oil was subjected to an international embargo. In an effort to bring Iranian oil production back to international markets, the U.S. State Department suggested the creation of a "Consortium" of major oil companies. The "Consortium for Iran" was subsequently formed by the following companies:

Anglo-Persian Oil Company (United Kingdom). This subsequently became Anglo-Iranian Oil Company and then British Petroleum. Following the acquisition of Amoco (which in turn was formerly Standard Oil of Indiana) and Atlantic Richfield it shortened its name to BP in 2000.

Gulf Oil (United States) In 1985 most of Gulf was acquired by Chevron, with smaller parts acquired by BP and Cumberland Farms. A network of service stations in the northeastern United States still bears the Gulf name.

The head of the Italian state oil company, Enrico Mattei sought membership for the Italian oil company Eni, but was rejected by what he dubbed the "Seven Sisters" - the Anglo-Saxon companies which largely controlled the Middle East's oil production after World War II. British writer Anthony Sampson took over the term when he wrote the book The Seven Sisters in 1975, to describe the shadowy oil cartel, which tried to eliminate competitors and control the world's oil resource.

Being well-organized and able to negotiate as a cartel, the Seven Sisters were initially able to exert considerable power over Third World oil producers. However, in recent decades the dominance of the Seven Sisters and their successor companies has been challenged by the increasing influence of the OPEC cartel (formed in 1960), a declining share of world oil & gas reserves held by OECD countries and the emergence of powerful state-owned oil companies in emerging-market economies.

Pops English

Don't Forget To Remember

Oh my heart won't believe that you have left me
I keep telling myself that it's true
I can get over anything you want my love
But I can't get myself over you

Don't forget to remember me
and the love that used to be
I still remember you
I love you
in my heart lies a memory
To tell the stars above
Don't forget to remember me my love

On my wall lies a photograph of you girl
Though I try to forget you somehow
You're the mirror of my soul
So take me out of my whole
Let me try to go on leavin' right now

Don't forget to remember me
and the love that used to be
I still remember you
I love you
in my heart lies a memory
To tell the stars above
Don't foeget to remember me my love

Success is never final. Failure is never fatal. It is courage that counts.

- Winston Churchill -

UNIT 11

How
Do Mooring Systems Work?

Mooring system은 어떻게 작용하는가?

Lecture A

How do Mooring Systems Work?

A mooring system is made up of a mooring line, anchor and connectors, and is used for station keeping of a ship or floating platform in all water depths. A mooring line connects an anchor on the seafloor to a floating structure. The mooring line can be made up of synthetic fiber rope, wire and chain or a combination of the three. Environmental factors - wind, waves and currents - determine which materials make up the mooring system. Chain is the most common choice for permanent moorings in shallow water up to 100 m, whereas steel wire rope is lighter weight and has a higher elasticity than chain, which is a better choice in water depths greater than 300 m. However, synthetic fiber rope is the lightest weight of all three. Configurations include all chain, chain and wire rope (conventional mooring line to 2,000 m), chain and synthetic fiber rope, and chain, wire rope and synthetic fiber rope combinations are used in ultra-deepwater (greater than 2,000 m). The mooring system relies on the strength of the anchors. The holding capacity of anchors depends on the digging depth and the soil properties.

The mooring lines run from the vessel to the anchors on the seafloor. Anchor types include: drag embedment, suction and vertical load. A drag embedment anchor (DEA) is the most utilized anchor for mooring floating MODUs in the Gulf of Mexico. The drag anchor is dragged along the seabed until it reaches the required depth. As it penetrates the seabed, it uses soil resistance to hold the anchor in place. The drag embedment anchor is mainly used for catenary moorings, where the mooring line arrives on the seabed horizontally. It does not perform well under vertical forces. Suction piles are the predominant mooring and foundation system used for deepwater development projects worldwide. Tubular piles are driven into the seabed and a pump sucks out the water from the top of the tubular, which pulls the pile further into the seabed. Suction piles can be used in sand, clay and mud soils, but not gravel, as water can flow through the ground during installation, making suction difficult. Once the pile is in position, the friction between the pile and the soil holds it in place. It can resist both vertical and horizontal forces. Vertical load anchors are similar to drag anchors as they are installed in the same way. However, the vertical load anchor can withstand both horizontal and vertical mooring forces. It is used primarily in taut leg mooring systems, where the mooring line arrives at an angle the seabed.

synthetic fiber 합성섬유 **elasticity** 탄성 the tendency of a body to return to its original shape after it has been stretched or compressed **strength** 강도 **suction** 흡입 a force over an area produced by a pressure difference **utilize** 이용하다 활용하다 **resistance** 저항 **predominant** 우월한 having superior power and influence **gravel** 자갈 **clay** 점토 **friction** 마찰 **withstand** 저항하다 잘 견디다 **soil** 흙 **seabed** 해저

There are six types of mooring systems discussed below. They include catenary, taut leg, semi-taut, spread, single point and dynamic positioning.

The catenary anchor leg mooring(CALM) system is the most commonly used system in shallow water. It gets its name from the shape of the free hanging line as its configuration changes due to vessel motions. At the seabed, the mooring line lies horizontally; thus the mooring line has to be longer than the water depth. Increasing the length of the mooring line also increases its weight. As the water depth increases, the weight of the line lessens the working payload of the vessel. In that case, synthetic ropes are used. As water depth increases, conventional, catenary systems become less and less economical.

The taut leg system typically uses polyester rope that is pre-tensioned until taut. The rope comes in at a 30 to 45 degree angle on the seabed where it meets the anchor (suction piles or vertically loaded anchors), which is loaded vertically. When the platform drifts horizontally with wind or current, the lines stretch and this sets up an opposing force. The semi-taut system combines taut lines and catenary lines in one system. It is ideally used in deepwater.

A spread mooring system is a group of mooring lines distributed over the bow and stern of the vessel to anchors on the seafloor. The vessel is positioned in a fixed heading, which is determined by the sea and weather conditions. The symmetrical arrangement of anchors helps to keep the ship on its fixed heading location. The spread mooring system does not allow the vessel to weathervane, which means to rotate in the horizontal plane due to wind, waves or current. Spread mooring is versatile as it can be used in any water depth, on any vessel, in an equally spread pattern or a group.

Single point mooring system connects all the lines to a single point. It links subsea manifolds connections and weathervaning tankers, which are free to rotate 360 degrees. The single point system includes a buoy, mooring and anchoring elements, product transfer system and other components.

Dynamic positioning does not use mooring lines. Instead a computer controls the vessel's thrusters and propellers to maintain position. DP can be used in combination with other mooring systems to provide additional redundancy.

catenary anchor leg mooring(CALM) a type of offshore mooring in which the facility is anchored by at least six anchors, it generally has no storage facility **oppose** 반대하다 **distributed** 분포된 spread out or scattered about or divided up **symmetrical arrangement** 대칭배열 **manifold** 많은 many and varied **transfer** 옮기다 the act of moving something from one location to another **additional** 부가의 **redundancy** 여분

Lecture B

After you read

1. What is the following is not a part of mooring system that is used for station keeping of a ship or floating platform in all water depths?
   (a) A mooring line
   (b) A anchor
   (c) Connectors
   (d) Thrusters

2. What is the most common choice for permanent moorings in shallow water up to 100 m?
   (a) Steel wire rope
   (b) Chain
   (c) Synthetic fiber rope
   (d) Combination of steel and synthetic fiber rope

3. Which is the better choice than the chain in water depths greater than 300 m?
   (a) Steel wire rope
   (b) Chain
   (c) Synthetic fiber rope
   (d) Combination of steel and synthetic fiber rope

4. What is not true about the article?
   (a) The catenary mooring system is the most commonly used system in deep water. .
   (b) The tout leg system typically uses polyester rope that is pre-tensioned until taut.
   (c) The semi-taut system combines taut lines and catenary lines in one system. It is ideally used in deepwater.
   (d) A spread mooring system is a group of mooring lines distributed over the bow and stern of the vessel to anchors on the seafloor.

Language Focus

1. Match the words with the definitions, then complete the definition with the given worlds.

The semi-taut system	a type of offshore mooring in which the facility is ________by at least six anchors
CALM	a group of mooring lines ________over the ________of the vessel to anchors on the seafloor
A spread mooring system	________taut lines and catenary lines in one system

bow and stern　　combines　　anchored　　distributed

2. Complete the sentences using the following words.

predominant　　seabed　　redundancy

1) Suction piles are the ________mooring and foundation system used for deepwater development projects
2) DP can be used in combination with other mooring systems to provide additional ____________.
3) The drag anchor is dragged along the _______until it reaches the required depth.

3. Complete the sentences using the following words.

1) A DEA is the most _______ anchor for mooring floating MODUs .
   a) utilized　　b) utilizing　　c) utility　d) utilization
2) The catenary anchor leg mooring system is the most ______ used system in shallow water.
   a) common　　b) commons　　c) commonly

Lecture D Useful Expressions- Expressions + ~ing

A. Be(get) used to + ~ing / used to ~ing

■ Bill is used to living alone. ☞ Bill has been living alone for 25years, so it is not strange for Bill to be home alone.

■ Bill used to live alone. ☞ Bill lived alone, but not anymore.

B. Have trouble[difficulty, a hard time, a problem] + ~ing

Tom had trouble making new friends. (*not* had trouble to make)

I had no difficulty finding a place to live in. (*not* no difficulty to find)

Did you have any trouble getting a visa? (*not* any trouble to get a visa)

C. go + ~ing for a number of activities, especially sports

go swimming, go hiking, go jogging, go fishing, go skiing, go shopping

How often do you go jogging?

I've never gone swimming. I am afraid of water.

D. It(There) is no use + ~ing. ☞There is nothing you can do about this situation.

It is no use crying over spilt milk.

= It is of use to cry over spilt milk.

= It is useless to cry over spilt milk.

D. be worth + ~ing

I live only a short walk from here. It is not worth taking a taxi.

What was the movie like? Was it worth watching?

This book is really good. It is worth reading.

Lecture E **Speak Out!!**

1. Talk with your partner. What would you say in these situation? Complete the sentences using the expressions you've studied in this unit.

past		current	
	A few years ago, Bill liked going outside, and he hated studying.		Now, he is a college student, so he has to study.

☞ A few years ago, Bill liked going outside, but he ________________ .
Now he is a college student. Studying was not easy for him but he finds it normal. He ____________________________________ .

2. Have a conversation with your partner based on the following example using Lecture D and practice twice. You can use your own ideas.

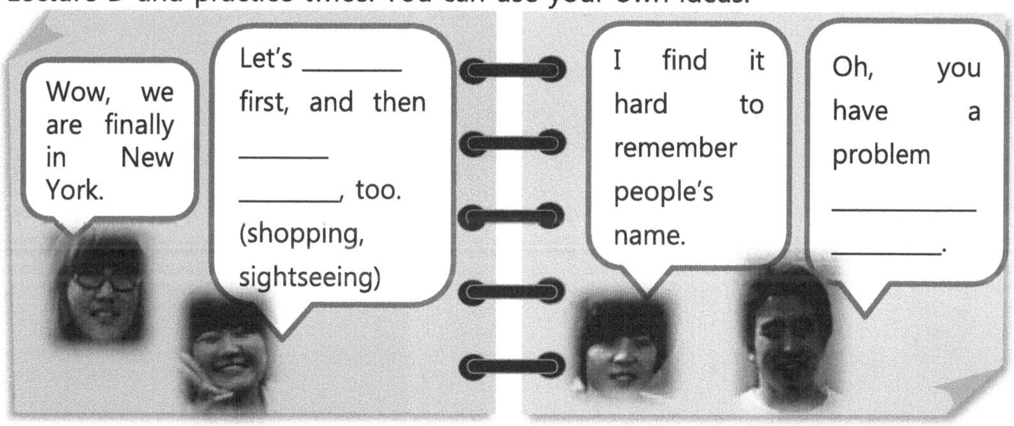

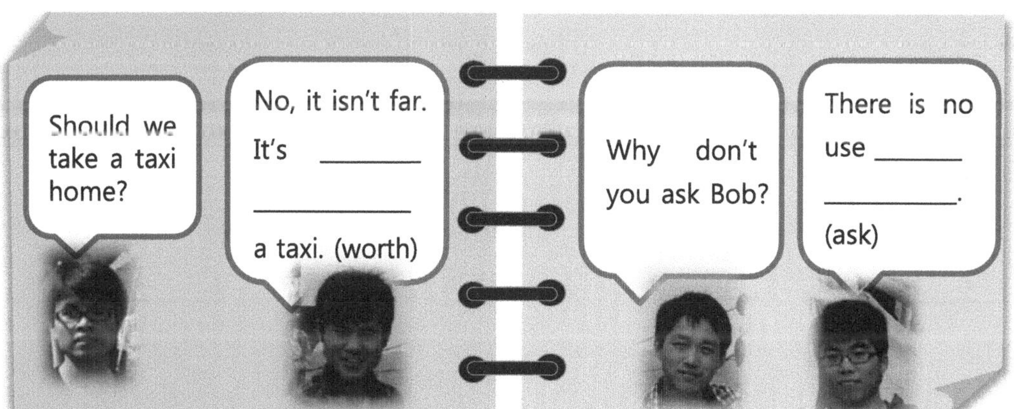

Lecture F Episode on offshore!!

Work in pair. Ask and answer the question.

	What are the materials that that can make up the mooring line?
	The mooring line can be made up of synthetic fiber rope, wire and chain or a combination of the three. Environmental factors – wind, waves and currents – determine which materials make up the mooring system.
	What is a drag embedment anchor?
	Answer the question!
	How can the tout leg system be used?
	Answer the question!
	*Make your own!*
	*Make your own!*

OUTRO The Organization of the Petroleum Exporting Countries

The Organization of the Petroleum Exporting Countries is an intergovernmental organization of twelve oil-producing countries made up of Algeria, Angola, Ecuador, Iran, Iraq, Kuwait, Libya, Nigeria, Qatar, Saudi Arabia, the United Arab Emirates, and Venezuela. OPEC has had its headquarters in Vienna since 1965 and hosts regular meetings among the oil ministers of its Member Countries. Indonesia withdrew in 2008 after it became a net importer of oil, but stated it would likely return if it became a net exporter again.

According to its statutes, one of the principal goals is the determination of the best means for safeguarding the organization's interests, individually and collectively. It also pursues ways and means of ensuring the stabilization of prices in international oil markets with a view to eliminating harmful and unnecessary fluctuations; giving due regard at all times to the interests of the producing nations and to the necessity of securing a steady income to the producing countries; an efficient and regular supply of petroleum to consuming nations, and a fair return on their capital to those investing in the petroleum industry.

OPEC's influence on the market has been widely criticized, since it became effective in determining production and prices. Arab members of OPEC alarmed the developed world when they used the "oil weapon" during the Yom Kippur War by implementing oil embargoes and initiating the 1973 oil crisis. Although largely political explanations for the timing and extent of the OPEC price increases are also valid, from OPEC's point of view, these changes were triggered largely by previous unilateral changes in the world financial system and the ensuing period of high inflation in both the developed and developing world. This explanation encompasses OPEC actions both before and after the outbreak of hostilities in October 1973, and concludes that "OPEC countries were only 'staying even' by dramatically raising the dollar price of oil.

OPEC's ability to control the price of oil has diminished somewhat since then, due to the subsequent discovery and development of large oil reserves in Alaska, the North Sea, Canada, the Gulf of Mexico, the opening up of Russia, and market modernization. As of November 2010, OPEC members collectively hold 79% of world crude oil reserves and 44% of the world's crude oil production capacity, affording them some control over the global market. The next largest group of producers, members of the OECD and the Post-Soviet states produced only 23.8% and 14.8%, respectively, of the world's total oil production.

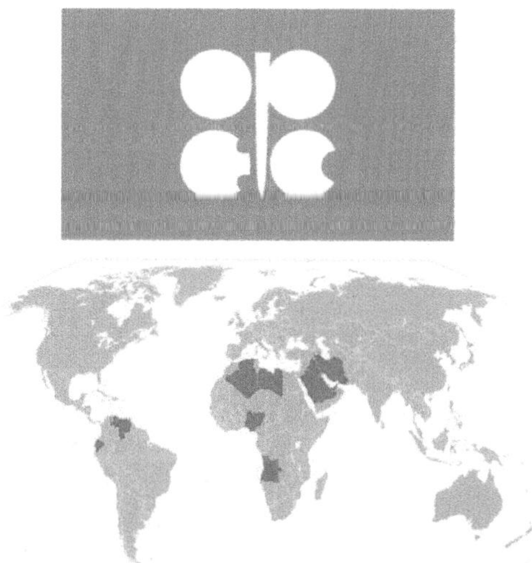

Pops English

You raise me up

When I am down and, oh my soul, so weary
When troubles come and my heart burdened be
Then, I am still and wait here in the silence
Until you come and sit awhile with me
You raise me up, so I can stand on mountains
You raise me up, to walk on stormy seas
I am strong, when I am on your shoulders
You raise me up, to more than I can be

You raise me up, so I can stand on mountains
You raise me up, to walk on stormy seas
I am strong, when I am on your shoulders
You raise me up, to more than I can be

There is no life - no life without its hunger;
Each restless heart beats so imperfectly;
But when you come and I am filled with wonder,
Sometimes, I think I glimpse eternity

You raise me up, so I can stand on mountains
You raise me up, to walk on stormy seas
And I am strong, when I am on your shoulders
You raise me up, to more than I can be

All our dreams can come true
if we have the courage to pursue
them.

- Walt Disney -

Unit 12

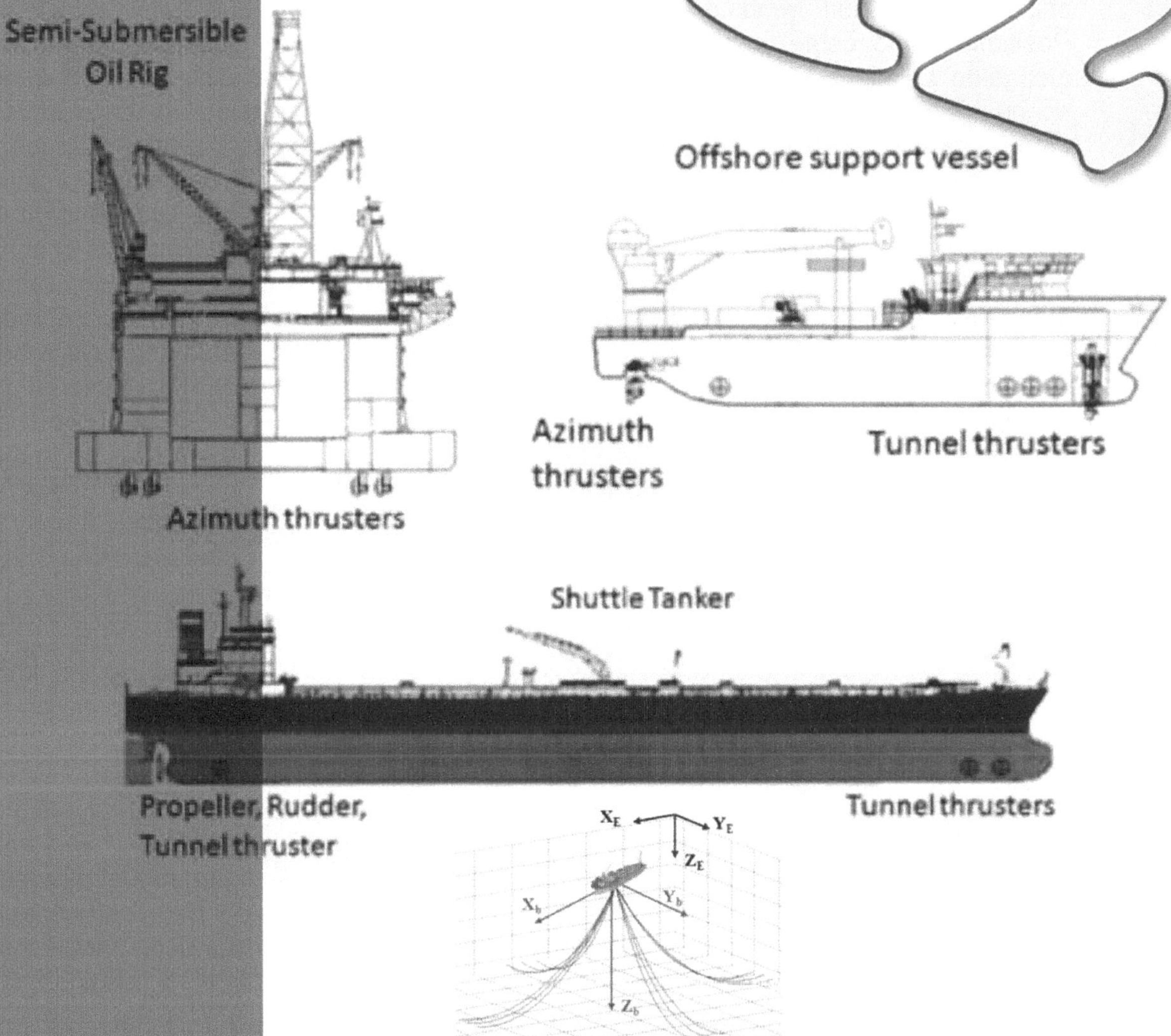

What

Is Dynamic Positioning?

DP는 무엇인가?

Lecture A

What Is Dynamic Positioning?

Dynamic positioning (DP) is a computer controlled system to automatically maintain a ship's position and heading by using her own propellers and thrusters. Dynamic Positioning has come a long way in the 58 years since its inception, developing alongside the oil industry. Many of today's advances in deep water exploration would not be possible without it, as it is pushed farther offshore to drill deeper in the quest for energy. A ship enabled with Dynamic Positioning can get to know about the changes in the wind and the waves and thus alter its course suitably without having to compromise on its main purpose. Ships enabled with Dynamic Positioning are independent of anchors and other support system in the sense that a Dynamic Positioning Ship enables the use of pushers and propellers to make the ship stay on course and steady rather than get carried away by the fluctuating winds and waves. This is perhaps the most advantageous feature of the system of Dynamic Positioning. In the earlier days, when ships used to enter the deeper parts of the ocean and the seas, there always used to be a threat of ships colliding with another ship because of natural movement of the wind and the waves or ships veering off course and thus getting lost and never to be found. But since the development of Dynamic Positioning which incorporates the usage of SONAR, Radar and other comprehensive detection, ships have started finding it easy to maintain their pace and steady their being in the deeper parts of the oceanic and sea waters.

As the 1950's came to an end, the Jack-up technology in use was reaching the limits of water depth they were able to drill in. Anchoring was becoming less economical. A new solution was needed. In 1961 the drill ship Cuss 1 (named for the Continental, Union, Shell & Superior oil consortium) was launched with 4 steerable propellers. While they were still operator controlled, the vessel did make use of Radar & Sonar ranging technology to determine position. Working within a radius of 180 m the Cuss 1 was able to drill a number of holes for the Mohole Project, the deepest being in 3500m of water. That same year Shell launched the Eureka. Equipped with thrusters fore & aft capable of rotating through 360 degrees, an analogue controller & a basic Taut Wire, known then as a Tilt Meter, she was the first true DP vessel.

In 1980 there were around 65 DP capable vessels operating in the offshore industry. With the integration of Satellite technology being used as a reference this number had increased to over 150 by 1985. GPS and its advances have paved the way, from that

compromise 타협하다 a middle way between two extremes **fluctuating** 변동을 거듭하는 having unpredictable ups and downs **enable** 가능하게 하다 render capable or able for some task **steady** 견고한 **economical** 경제적인 using the minimum of time or resources necessary for effectiveness **determine** 결정하다 **integration** 통합 **threat** 협박 **solution** 해결책

time until the present, for the growth of the worldwide DP fleet. While the early vessels equipped with a satellite reference system were primarily concerned with static station keeping, with time and experience this eventually moved to include precise vessel movement. More recent advances in satellite & ground based augmentation systems such as DGPS & WAAS have reduced position error to less than half a meter, enabling ships to perform the precision operations. Tasks such as pipe lay are now performed with great accuracy, allowing optimal use of resources.

As DP continued to be utilized through the 80′s and into the 90′s it was evident that regulation was required due to a lack in consistency in how DP control & associated issues were managed. Consultants had been working together informally through the 80′s to create procedures to cope with this new form of vessel handling & by 1990 the Dynamic Positioning Vessel Owners Association (DPVOA) was formed. In 1995 the DVPOA merged with the International Association of Diving Contractors (AODC) to form the International Marine Contractors Association (IMCA), which remains the main DP regulatory body to this day. Since then IMCA has released over 50 guidance documents, based on the lessons learned and recorded over the years. Moving now towards 21 century and beyond, DP is an integral part of the offshore oilfield. Advances in technology & reference systems make the systems more reliable & greatly increase the Mean Time Between Failure (MTBF).

Today it has being drilled in deeper water; farther offshore then could be imagined back on the Eureka. Innovative references such as a deep water taut wire constructed from fiber optic cable that can measure the curve due to current allows vessels such as the newly built Discoverer Clear Leader, Chevrons newest Enterprise Class drill ship, to work in 3650m of water, drilling to a total depth of 12,200 m. These advances can be largely attributed to Dynamic Positioning technology.

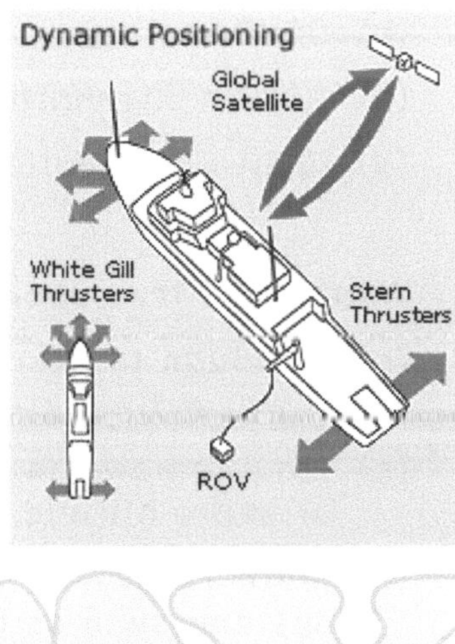

reference 언급, 참조 **static** 정적인 not in physical motion **augmentation** 증대, 증가 the amount by which something increases **accuracy** 정확성 the quality of being near to the true value **optimal** 최상의 most desirable possible under a restriction expressed or implied **resource** 자원 **associated** 연합의 **innovative** 혁신적인 **attributed** 결과라고 생각한다

Lecture B

After you read

1. When was the Jack-up technology in use reaching the limits of water depth they were able to drill in?
   (a) The end of 1950'
   (b) The beginning of 1950'
   (c) The mid of 1950'
   (d) In 1950

2. What is the name of the first true drill ship which was equipped with thrusters fore & aft capable of rotating through 360 degrees, an analogue controller & a basic Taut Wire?
   (a) Eureka
   (b) The drill ship Cuss 1
   (c) DPVOA
   (d) IMCA

3. Fill in the blanks.
   (a) MTBF : Mean Time Between ______
   (b) DPVOA : _______ Positioning Vessel Owners Association
   (c) AODC : International Association of _____________
   (d) IMCA : International Marine ________ Association

4. What is not true about the article?
   (a) Although the Eureka ushered in a new age of offshore technology it was mainly used for drilling core samples.
   (b) Moving now towards 2010 & beyond, DP is an integral part of the offshore oilfield.
   (c) More recent advances in satellite & ground based augmentation systems such as DGPS & WAAS have reduced position error to less than 5 meters.
   (d) Tasks such as pipe lay are now performed with great accuracy, allowing optimal use of resources.

Lecture c	Language Focus

1. Match the words with the definitions, then complete the definition with the given worlds.

Word		Definition
DP	●	a computer ________ system to automatically ________a ship's position and heading by using her own propellers and thrusters.
economical	●	having ________ups and downs
fluctuating	●	using the minimum of time or resources necessary for effectiveness

maintain effectiveness unpredictable controlled

2. Complete the sentences using the following words.

threat optimal accuracy enable

1) Ships ________with Dynamic Positioning are independent of anchors and other support system.
2) There always used to be a ______ of ships colliding with another ship.
3) Tasks such as pipe lay are now performed with great ______, allowing ________ use of resources.

3. Complete the sentences using the following words.

1) In 1980 there were 65 DP capable vessels operating in the offshore industry.
   a) around b) on c) in d) for

Lecture D — Useful Expressions – If I were you

1. If I knew~/ I wish I knew

If + S + *past* (V)~, S + would/could/should/might + V	or	I wish I *past* (V)~

When you imagine a present situation which is impossible, you can use this structure.

If I knew Tom's number, I would call him.	I wish I knew Tom's number	☞ I don't know Tom's number, so I regret this.
It I were a bird, I could fly to you.	I wish I were a bird.	☞ I can't fly.
If I won a million dollars, I would faint.	I wish I won a million dollars.	☞ I can't win a million dollars.

2. If I had known~/ I wish I had known~

If S + had P.P~, S + would/could/should/might + had P.P	or	I wish I had P.P~

When you imagine a past situation which was impossible, you can use this structure.

If I had known Tom's illness, I would have gone to see him.	I wish I had known Tom's illness. I would have gone to see him.	☞ I didn't know Tom's illness.
It I were a bird, I could fly to you.	I wish I were a bird.	☞ I can't fly.
If I had won a million dollars, I would have sent many poor students to college.	I wish I had won a million dollars.	☞ I couldn't win a million dollasr.
I shouldn't have done this!!	I wish I hadn't done this.	☞ I did this, but I regret this.

Speak Out!!

1. Talk with your partner. What would you say in these situation? Complete the sentences using the expressions you've studied in this unit.

2012. 05.13

Today, it is such nice weather.
The sky is so blue, and everything is so beautiful.
Suddenly, I want to be a bird!! What a good dream!

Another hope!
I want a girlfriend!!!

I wish ______________________

____

If I ______________________,

I ______________________.

I wish ______________________.

2012. 06.30

Today is my exam day.
Oh my God.
I didn' t study at all.

I regret I played basketball yesterday.

I regret I slept early last night.

I wish ______________________

____

If I ______________________,

I ______________________.

I wish ______________________.

Lecture F   Episode on offshore!!

Work in pair. Ask and answer the question.

	What is DP?
	Dynamic positioning (DP) is a computer controlled system to automatically maintain a ship's position and heading by using her own propellers and thrusters.
	Why is Dynamic positioning used?
	Answer the question!
	What is the most advantageous feature of the DP?
	Answer the question!
	*Make your own!*
	*Make your own!*

Marine Mineral

The oceans hold a veritable treasure trove of valuable resources. Sand and gravel, oil and gas have been extracted from the sea for many years. In addition, minerals transported by erosion from the continents to the coastal areas are mined from the shallow shelf and beach areas. These include diamonds off the coasts of South Africa and Namibia as well as deposits of tin, titanium and gold along the shores of Africa, Asia and South America.

Efforts to expand ocean mining into deep-sea waters have recently begun. The major focus is on mangane-se nodules, which are usually located at depths below 4000 metres, gas hydrates (located between 350 and 5000 metres), and cobalt crusts along the flanks of undersea mountain ranges (between 1000 and 3000 metres), as well as massive sulphides and the sulphide muds that form in areas of volcanic activity near the plate boundaries, at depths of 500 to 4000 metres.

Back in the early 1980s there was great commercial interest in manganese nodules and cobalt crusts. This initial euphoria over marine mining led to the International Seabed Authority (ISA) being established in Jamaica, and the United Nations Convention on the Law of the Sea (UNCLOS) being signed in 1982 – the "constitution for the seas". Since entering into force in 1994, this major convention has formed the basis for signatories' legal rights to use the marine resources on the sea floor outside national territorial waters (Chapter 10).After that, however, the industrial countries lost interest in resources. For one thing, prices dropped – making it no longer profitable to retrieve the accretions from the deep sea and utilize the metals they contained. Also, new onshore deposits were discovered, which were cheaper to exploit. The present resurgence of interest is due to the sharp increase in resource prices and attendant rise in profitability of the exploration business, and in particular to strong economic growth in countries like China and India which purchase large quantities of metal on world markets.

Even the latest economic crisis is not expected to slow this trend for long. The industrial and emerging countries' geopolitical interests in safeguarding their supplies of resources also play a role. In light of the increasing demand for resources, those countries which have no reserves of their own are seeking to assert extraterritorial claims in the oceans.

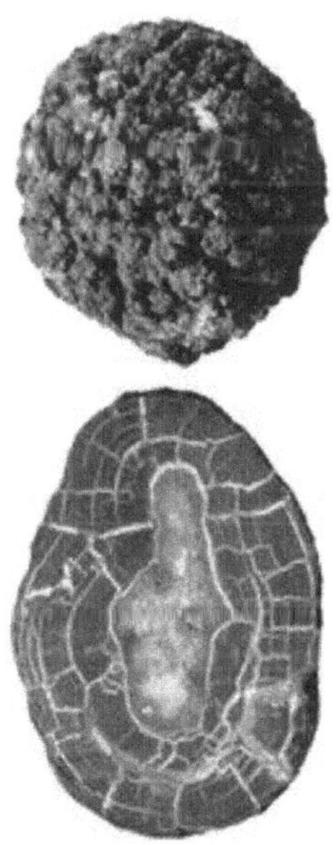

Cross-section view of a manganese nodule: Over millions of years, minerals are deposited around a nucleus.

Pops English

Love Of My Life

Love of my life
You've hurt me
You've broken my heart
And now you leave me
Love of my life
Can't you see
Bring it back, bring it back
Don't take it away from me
Because you don't know what it means to me

Love of my life
Don't leave me
You've taken my love, you now desert me
Love of my life
Can't you see
Bring it back, bring it back
Don't take it away from me
Because you don't know what it means to me

You will remember when this is blown over
And everything's all by the way -
When I grow older
I will be there at your side
To remind you how I still love you
I still love you

Back, hurry back
Please bring it back home to me
Because you don't know what it means to me
Love of my life
Love of my life

The greatest risk is the risk of riskless living

-Steven Corby

UNIT 13

What
Is The Turret Mooring system?

Turret mooring system은 무엇인가?

Lecture A

What Is The Turret Mooring System?

The turret is the major component of a vessel or any floating production unit (FPSO, FPO) to which the risers coming from the seabed are connected. Turret name refers the rotating function of this large equipment to facilitate the rotation of the vessel according to winds, waves and dominating current while the risers remain connected to a fix part on the seabed. In relatively calm waters, such as in West Africa, turrets can be located externally to the ship structure, hanging off the bow of the FPSO. For harsher environments like the North Sea, the turret is generally located internally.

The turrets, mooring lines and risers can be designed to be permanently connected in order to remain on location for long periods of time and disconnectable, capable of quick disconnection through a procedure lasting just a few hours. Most ship-shaped FPSOs in the North Sea are purpose-built and are permanently moored. In areas of the world subject to cyclones (northwestern Australia) or icebergs (Canada), some FPSOs are able to release their mooring/risers turret and steam away to safety in an emergency. The turret sinks beneath the waves and can be reconnected later.

The turret fulfils more functions than weathervaning. It also uplift the oil and gas and associated water and condensates production risers as well as it supports piping for water or gas injection when used. The turret supports the (multi pass) fluid swivel and forms the interface between sub-sea facilities and topsides. The electrical power supply of submerged equipment, the control system, all the information and communication systems pass also through the turret. As all risers pass through the inside of the turret bearing, it is considered as the bottleneck to the number of risers and therefore potential further field production enhancement. With a turret the offloading can be executed in tandem to the FPSO with shuttle tankers. This provides a more operational uptime and less risk of tanker collision compared with spread-moored FPSOs. An FPSO turret requires a highly specialized fabrication. If the turret is attached to the floating unit, the turret is part of the mooring system and called Turret Mooring system.

component 구성요소 **refer** 언급하다 **relatively** 상대적으로 **procedure** 절차 **uplift** 들다, 올리다 **condensate 응축액** **bottleneck** 좁은통로 a narrowing that reduces the flow through a channel **swivel** 회전 a coupling (as in a chain) that has one end that turns on a headed pin **execute** 실행하다 **uptime** 가동시간 **fabrication** 조립, 구성 **enhancement** 향상 an improvement that makes something more agreeable

Depending on the type of vessel its operations location internal or external turret may be used. The internal turret is located in the front end of a vessel. Sometimes it can be found in the middle. As the internal turret is appropriate for a large number of risers, it has good fluid transfer capabilities. The largest internal systems can accommodate up to 100 or more risers. Internal turret mooring systems also allow vessels to remain on location permanently. This holds even in the most harsh environments, as can be seen with the FPSOs in the North Sea. An advantage of the internal turret in comparison to the external turret is that mooring forces can be transferred more easily into the hull.

The external turret is located outside the ship's hull. It comprises a steal box located at the bow or stern of the vessel, which provides a foundation for the bearing and turret. The external turret is mostly set-up above the waterline, but sometimes also under. A swivel provides product and utility connections between the tanker facilities and the seabed. Being positioned outside of the hull the external turret can bear less risers and production equipment than internal turret. But the design of the external turret make it easier to be mounted on tankers converted into FPSO. So most of the new-built or purpose-built FPSO for large oil and gas fields will use internal turret while converted FPSO shall be preferred for medium or smaller developments.

GLOSSARY

appropriate 적합한 **permanently** 영구히 for a long time without essential change **convert** 바꾸다 change from one system to another or to a new plan or policy **prefer** 하기를 좋아한다 **internal** 내부의 **external** 외부의 coming from the outside **bear** 견디다 put up with something

Lecture B

After you read

1. What is the major component of a vessel or any floating production unit (FPSO, FPO) to which the risers coming from the seabed are connected?
   (a) Spread mooring
   (b) Turret
   (c) Bow loading system
   (d) DP system

2. What is not the function of turret?
   (a) To weather-vane.
   (b) To support piping.
   (c) To uplift production risers.
   (d) To discharge gas.

3. Please complete the sentence.
   (a) Turret name refers the ________function of this large equipment to facilitate the rotation of the vessel according to winds, waves and dominating current while the ________remain connected to a fix part on the seabed.
   (b) In relatively calm waters, such as in West Africa, turrets can be located _______to the ship structure, hanging off the bow of the FPSO.
   (c) If the turret is attached to the floating unit, the turret is part of the mooring system and called____________________________.

4. What is an advantage of the internal turret in comparison to the external turret?
   (a) Mooring forces can be transferred more easily into the hull.
   (b) Internal turrets require less mooring forces.
   (c) Internal turrets have the simple structure.
   (d) Mooring forces can be partial transferred into the hull

Lecture c	Language Focus

1. Match the words with the definitions, then complete the definition with the given worlds.

Words	Definitions
enhancement	the major ________of a vessel or any floating production unit to which the risers ________the seabed are connected
turret	___________something/someone
bear	an ____________that makes something more agreeable

improvement   coming from   component   put up with

2. Complete the sentences using the following words.

external   execute   internal

1) The ___________turret is located in the front end of a vessel.
2) The _____________turret is located outside the ship' s hull.
3) With a turret the offloading can be _____________in tandem to the FPSO with shuttle tankers.

3. Complete the sentences using the following words.

1) This provides a more operational uptime and less risk of tanker collision ____________ with spread-moored FPSOs.

a)filled   b) compared   c) completed

2) An FPSO turret requires a __________ specialized fabrication.

a) highly   b) high   c) height

Lecture D **Useful Expressions- how to do**

1. how to + R

■ You can learn **how to skate** at the ice rink.
= You can learn **how you should skate** at the ice rink.

■Through his other books, Max taught us **how to** live peacefully.
=Through his other books, Max taught us **how we should live** peacefully.

2. what to + R

■ I don't know **what to do** for my girlfriend's birthday.
= I don't know **what I should do** for my girlfriend's birthday.

■ We usually don't have problems when deciding **what to eat**.
=We usually don't have problems when deciding **what we should eat**.

3. where to + R

■ We haven't decided **where to stay** during our trip.
= We haven't decided **where we should stay** during our trip.

■ He didn't explain **where to live** after moving out.
=He didn't explain **where he should live** after moving out.

4. when to + R

■ You are learning **what to watch** for and **when to stop**.
= You are learning **what you should watch** for and **when you should stop**.

■ Mondale is a politician who knows **when to keep** his mouth shut.
=Mondale is a politician who knows **when he should keep** his mouth shut.

■ Complete sentences using words from the box below.

play wear turn

1. I can't decide ____________________ for the party.
2. Does Bill know ____________________ golf?
3. My mother didn't tell me ________________ off the oven.

Lecture E	Speak Out!!

A. Talk with your partner. What would you say in these situation? Complete the sentences using the expressions you've studied in this unit.

	You need to learn ____________ your cell phones wisely! (use)
	Now she also teaches people ______________themselves from many diseases. (protect)
	I usually don't have problems when deciding __________ but in this restaurant everything looks really good. I can't decide ______________!

2. Have a conversation with your partner based on the following example using Lecture D and practice twice. You can use your own ideas.

Ex.

	I am in trouble. My car just broke down. I don't know <u>what to </u>do.		Ok. calm down. Firstly, You can check it by yourself, such as fuel, battery, etc.
	I think everything looks good. I have no idea <u>how to</u> fix the car.		You don't have to do that. Only what you have to do is to call the car garage.
	But I don't know <u>where to</u> call.		It's ok. I have one.

Situation 1	Situation 2
You lost your wallet. Firstly, you look everywhere around you. Then, you call the police.	*Make your own!*

Episode on offshore!!

Work in pair. Ask and answer the question.

	What is the turret?
	Turret is the major component of a vessel or any floating production unit (FPSO, FPO) to which the risers coming from the seabed are connected.
	What is the main function of the turret ?
	Answer the question!
	What is an advantage of the internal turret in comparison to the external turret?
	Answer the question!
	*Make your own!*
	*Make your own!*

OUTRO

Tidal Energy & Current Energy

Tidal power plants work in a similar way to power plants at a reservoir – except that the water masses do not flow downhill but are moved back and forth with tidal flows. Unlike other forms of ocean energy, tidal energy has been utilized commercially for some time. The La Rance tidal power station began operations in 1966 at St. Malo on the Atlantic coast in northern France, where the La Rance River flows into the sea. At high tide the water flows upstream through the large turbines of the power station, and at low tide it flows downstream again. The 240 megawatt (MW) power station(1 megawatt = 1 million watts) has a similar output to a gas-fired power station. Similar facilities have been constructed in Canada, China and Russia over the past 20 years, although these are considerably smaller. A 260 MW tidal power station integrated in an existing dam is scheduled to come on-stream in South Korea this year.

The United Kingdom has been planning to construct a major tidal power station at the estuary of the River Severn between England and Wales for some time. The location could supply enough energy to meet 7 per cent of the United Kingdom's entire power needs. However, critics fear that the construction of the dams could devastate vital nature reserves and bird sanctuaries. The environmental damage could be substantial. For this reason alternative concepts and locations are now being discussed.

Ocean current energy can also be harnessed using submerged rotors which are driven by the motion of the water.It has been estimated that ocean current power stations and tidal power plants together could harness several 100 terawatt-hours of electricity per year worldwide.

For some time now tests have been carried out on some rotor concepts, such as the Seaflow system, the prototype of which commenced operations off the English coast in 2003. Its successor, SeaGen, is now rotating in the Strangford Narrows off the Irish coast. Under this concept two rotors are mounted on the tower of the plant. This increases the electricity yield and balances out the high construction and start-up costs.

Such ocean facilities face much harsher stresses from currents and wave movements than wind turbines, for example, and for this reason extensive endurance testing is called for. Nonetheless, the SeaGen technology is closely based on the wind turbine model. The blade angle and rotational speed can be adjusted to suit the prevailing current. Other concepts focus on fixed, non-adjustable systems.

Pops English

I have a dream

I have a dream, a song to sing
To help me cope with anything
If you see the wonder of a fairy tale
You can take the future even if you fail

I believe in angels
Something good in everything I see
I believe in angels
When I know the time is right for me
I'll cross the stream - I have a dream

I have a dream, a fantasy
To help me through reality
And my destination makes it worth the while
Pushing through the darkness still another mile

I believe in angels
Something good in everything I see
I believe in angels
When I know the time is right for me
I'll cross the stream - I have a dream
I'll cross the stream - I have a dream

I have a dream, a song to sing
To help me cope with anything
If you see the wonder of a fairy tale
You can take the future even if you fail

I believe in angels
Something good in everything I see
I believe in angels
When I know the time is right for me
I'll cross the stream - I have a dream
I'll cross the stream - I have a dream

And in the end, it's not the years in your life that count. It's the life in your years.

- Abraham Lincoln

UNIT 14

What
Is a LNG Re-gasification Vessel?

LNG-RV는 무엇인가?

Lecture A What Is LNG Re-gasification Vessel?

In August 2005, Hurricane Katrina swept the eastern United States, severing power lines and making all appliances in the area useless. The big question was how to supply energy to desperate New Orleans residents. One solution was shipping liquefied natural gas (LNG) to a refinery. But the Katrina had left all of New Orleans' processing plants unusable. When hurricane Katrina destroyed most of the gas facilities near the Gulf of Mexico, only the LNG-RVs (Regasification Vessel) the Korean shipbuilder made remained intact. An LNG RV proved the best solution since it could process the liquefied gas on the ship. Despite the still formidable waves, this ship stably carried energy and allowed New Orleans residents to turn on the power again. This ship was the brainchild of a U.S. firm Accelerator Energy and Exmar in early 2000, which chose Daewoo as its builder. When Daewoo was asked if it could build such a ship, the firm did not hesitate to answer "yes." The 3 parties have worked very closely for the LNG RV project from the early concept development to the detail engineering stages. Though the concepts and ideas on the offshore floating LNG terminal have been studied and widely spread for many years, actual projects could not be materialized mainly due to technical risks involved in the offshore operation as well as the commercial justification. Considering land-based or offshore LNG receiving terminal projects usually take 5-7 years from feasibility study, authority permission clearance, engineering, and actual construction, the first physical appearance of the LNG RV in the commercial market would be an important achievement in the offshore LNG technology.

LNG RV is a vessel with combined functions of the conventional LNG carrier, offshore mooring turret and regasification (regas) facilities. As the vessel has onboard regas facility, the regasified natural gas (NG) can be connected directly to a commercial pipeline trunk. Therefore the process does not need any land-based LNG receiving and regas facilities, also it can save construction costs for a ground terminal and make facilities less vulnerable to terrorist attacks. The LNG RV concept is also known as the "Energy Bridge". Though the LNG RV concept is unique, the technologies used in LNG RV are based on proven conventional ones. The long operation records of the ocean-going LNG vessels have proved the reliabilities of the onboard equipment and the LNG containment system. The similar onboard turret systems have been used for many years for crude oil shuttle tanker applications.

severe 심한, 엄한 **sweep** 휩쓸고지나가다 force into some kind of situation, condition, or course of action **intact** 온전한 **formidable** 만만찮은 **resident** 거주자 **hesitate** 주저하다 **permission** 허락, 허가 **justification** 정당화 **materialize** 유형화하다 **feasibility** 실현가능성 the quality of being doable **vulnerable** 공격당하기 쉬운 susceptible to attack **attack** 공격 **reliability** 신뢰도 the quality of being dependable or reliable **application** 적용

The typical LNG storage and regasification(regas) capacity of the LNG RV is 138,000 M3 and 500 mmscfd respectively. The LNG RV had successfully completed the submerged buoy connection test and the regas operation test with actual LNG during the sea and regas trials before the delivery of the vessel. The LNG RV is based on the conventional LNG carrier design with onboard LNG regasification (regas) facilities and internal turret for the sub-sea pipe connection. The LNG RV can take the role of the conventional LNG carrier during the voyage, and at the same time it is used for an offshore regas terminal when connected to a submerged buoy.

The regas process is mainly composed of the LNG feeding pumps, HP (high pressure) pumps, vaporizers, and send-out equipment, which is similar to that of a land-based LNG receiving terminal. The heating medium of the vaporization is sea water. The LNG cargo tanks are based on the conventional membrane type (GTT No. 96) containment system. But the strengthened insulation boxes have been used in sides, hoppers and top of the tanks. Therefore, there is no tank level restriction in the regas operation. In LNG RV, 6 sets of HP pumps and vaporizers are installed in the forward part of the vessel (3 units in port, 3 units in starboard) The normal regasification capacity is 500 mmscfd (each vaporizer : 100 mmscfd, million standard cubic feet per day), and the maximum capacity can be 690 mmscfd. Therefore, the onboard 138,000 cubic meters LNG can be converted to the HP NG within 5-6 days. LNG RV process is already shown in Unit 9. The HP NG also can be delivered through either mid port or starboard HP manifold.

feed 먹이다,공급하다 feed into; supply **vaporizers** 기화기 a device that puts out a substance in the form of a vapor (especially for medicinal inhalation **restriction** 제한 an act of limiting or restricting **briefly** 간단히

Lecture B **After you read**

1. What is not the regas process composed of?

(a) LNG feeding pumps

(b) vaporizers

(c) send-out equipment

(d) Low pressure pump

2. What does the LNG RV look like?

(a) Conventional LNG carrier with onboard LNG regas facilities

(b) Conventional LNG carrier with onboard LNG requefaction facility

(c) Floating storage facility with onboard LNG regas facilities.

(d) Floating storage facility with onboard LNG requefaction facility.

3. What is also RV known as ?

(a) Energy bridge

(b) Energy circle

(c) Energy backup.

(d) Energy recycle

4. Please complete the sentence.

(a) The LNG RV can take the role of the conventional LNG carrier during the _______, and at the same time it is used for an _________ _________________when connected to a submerged buoy.

(b) The heating medium of the vaporization is ____ ______.

(c) The strengthened insulation boxes have been used in sides, ___________ and top of the _________.

Language Focus

1. Match the words with the definitions, then complete the definition with the given worlds.

Words	Definitions
LNG-RV	the quality of being ________ or reliable
feasibility	the quality of being ________
reliability	a vessel with ________functions of the conventional LNG carrier, offshore mooring turret and ____________facilities.

doable   combined   regasification   dependable

2. Complete the sentences using the following words.

justification   residents   materialize   formidable

1) Despite the still _______waves, this ship stably carried energy and allowed New Orleans ________ to turn on the power again.
2) Actual projects could not be __________mainly due to technical risks involved in the offshore operation as well as the commercial ____________.

3. Complete the sentences using the following words.

1) __________the LNG RV concept is unique, the technologies used in LNG RV are based on proven conventional ones.
   a) When   b) As   c) Because   d) Though
2) The LNG RV is based ___ the conventional LNG carrier design.
   a) on   b) to   c) with   d) at

Lecture D Useful Expressions- Prepositions

1. above, below, over, under

- **above** The airplane was flying above the clouds.
- **below** There is a vase below the portrait.
- **over** Hundreds of birds passed over my head.
- **under** I found that my cat was hiding under the bed.

2. out of, onto , up, down, into, off

- **out of** They got out of the car after two hours.
- **into** We went into the forest to find the hidden treasure.
- **up** Mark ran up the stairs from the first floor.
- **down** Go down one floor and turn left.
- **onto** The cat jumped onto the chair.
- **off** I once fell off the top of the ladder.

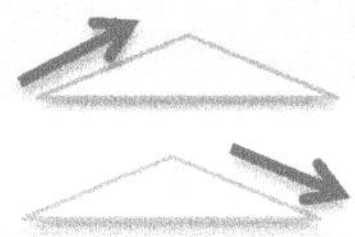

3. Complete sentences using words from the box below.

play wear when how turn what go where

1. I can't decide __________________ for the party.
2. Does Bill know ____________________ golf?
3. My mother didn't tell me ________________ off the oven.
4. I don't know __________________ , I think I am lost.

Lecture E **Speak Out!!**

1. Talk with your partner. What would you say in these situation? Complete the sentences using the expressions you've studied in this unit and match each other.

a.

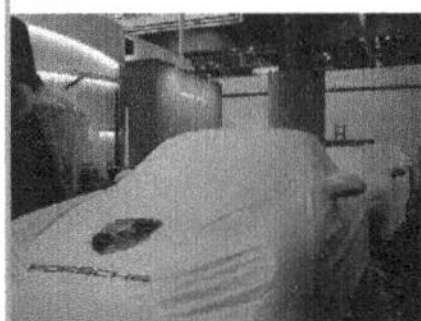

b.

c.

d.

1) The sun is dropping ________ the horizon.	
2) The salesman put a grey cover ________ the new car.	
3) A bird is flying just ______ the tree tops.	
4) The rabbit was _____ the desk.	

2. Complete the sentences using the expressions you've studied in this unit and match each other.

a

b

c

d

1) An astronaut is coming ______the spaceship. He climbs _____ a ladder ______ the surface of the planet.

2) Only a small part of an iceberg shows ______ the surface of the sea. Most of it lies _____ the surface where we cannot see it.

3) ______ the sea, colorful fish swim together. Hundreds of birds are flying ______ the sea.

4) The bottle fell ______ the table. The powder was spilt ______ the floor.

Lecture F **Episode on offshore!!**

Work in pair. Ask and answer the question.

	What happened in August 2005?
	In August 2005, Hurricane Katrina swept the eastern United States, severing power lines and making all appliances in the area useless.
	What is a LNG-RV?
	Answer the question!
	What is the advantage of LNG-RV?
	Answer the question!
	Make your own!
	Make your own!

Six Day War and advent of VLCC

The war was one of the shortest in history but created major disruption to shipping. The Suez Canal was closed for eight years, forcing operators to change their routes and commercial strategies. The canal, connecting the Mediterranean Sea and the Red Sea, opened in 1869 and slashed journey times between Europe, the East and Australasia.

The Six Day War and the 1973 Arab-Israeli conflict resulted in an Egyptian blockade of the canal and shipping lines assumed correctly it would remain closed for a very long time. The huge bulk oil tanker Titan was one of many Very Large Crude Carriers (VLCCs) designed during this period when operators knew they could not use Suez. They were too big to go through the canal but their large size made them more cost-effective for travelling the extra distances. Oil transportation was one of the most profitable shipping sectors at the time. When OPEC (the Organisation of Petroleum Exporting Countries) quadrupled oil prices in 1973 it triggered a worldwide slump in shipping.

Titan was built in 1970 in Gothenburg, Sweden, and registered in Liverpool with the famous Blue Funnel Line (Ocean Steam Ship Company). There is a superb six-foot long model of the 113,551- ton tanker on display in Merseyside Maritime Museum (pictured). Titan only sailed under Blue Funnel colours for five years before being sold to Mobil Oil in 1975. Just seven years later she was sold for scrap in South Korea.

By 1982, when there were 577 VLCCs in the world, it was found that 326 of them including Titan were surplus to requirements. Photographs show other VLCCs of the era including a deck view of BP tanker British Admiral about 1970. The main engine room of the British Mariner shows crew members dwarfed by enormous pipes and machinery. Titan was the fourth and last Blue Funnel ship to bear that name. The first Titan was built in 1885 by Scott & Co of Greenock and broken up in 1902.

The second Titan, built in 1906, was torpedoed and sunk in 1940 by the German submarine U-47 with the loss of six lives. The U-boat was commanded by Günther Prien, a notorious ace who sank more than 30 Allied ships including the veteran British battleship Royal Oak. Titan was the 18th vessel he sent to the bottom.

This is an edited version of the Maritime Tale that originally appeared in the Liverpool Echo.

Pops English

I will follow him

I will follow him
Follow him wherever he may go
And near him I always will be
For nothing can keep me away
He is my destiny

I will follow him
Ever since he touched my heart I knew
There isn't an ocean too deep
A moutain so high it can keep
Keep me away
Away from his love

[Moderately Fast Rock]
I love him, I love him, I love him
And where he goes I'll follow, I'll follow, I'll follow
I will follow him
Follow him wherever he may go
There isn't an ocean too deep
A moutain so high it can keep
Keep me away

We will follow him
Follow him wherever he may go
There isn't an ocean too deep
A mountain so high it can keep
Keep us away
Away from his love

(I love him) Oh yes I love him
(I'll follow) I'm gonna follow
True love, he'll always be my true love
(Forever) From now until forever
I love him, I love him, I love him
And where he goes I'll follow, I'll follow, I'll follow
He'll always be my true love, my true love, my true love
From now until forever, forever, forever
There isn't an ocean too deep
A mountain so high it can keep
Keep us away
Away from his love

I think death is the most wonderful
invention of life, because we're
completely aware of it and we
make it a priority
lf says I'm possible...

- Steve Jobs

UNIT 15

How Do ROVs Work?

왜 ROV가 필요한가?

Lecture A

How Do ROVs Work?

Remotely Operated Vehicles, or ROVs, are underwater robots that allow the controller to be located above the water. ROVs are connected to the topside via an umbilical link that houses communications cables, an energy source and information transfer. All ROVs contain some sort of visual device, such as a camera, to see under the water, and additional tools vary according to what type of ROV is being used. For example, some ROVs have multiple manipulators to carry out myriad operational functions at water depths up to 10,000 feet. As ROV technologies and capabilities have improved, the offshore oil and gas industry has come to rely on ROVs for drilling support and subsea construction services to enable deepwater exploration and development projects worldwide. Human divers can only submerge to certain depths, and the skill and danger involved makes deepwater diving unrealistic. Developed to overcome the limitations of human subsea divers, ROVs have become an indispensable tool to the offshore oil and gas industry.

The first iteration of the robotic diving concept was the Human Occupied Vehicle, or HOV, which was developed in the mid-1960s. While somewhat successful, HOVs were limited by human frailties and unable to achieve the water depths needed. In 1966, the US Navy recovered an atomic bomb lost in a plane crash offshore Spain using an ROV, and in 1973, an ROV was used to save sailors in a sunken submersible offshore Ireland.

The commercial world quickly began manufacturing preliminary versions of ROVs for the offshore oil and gas industry. Although these products were not as successful initially, the technology has come a long way, and now ROVs are regularly used offshore for petroleum developments and drilling, military and science efforts, as well as underwater cable endeavors. ROVs can vary in size from that of a bread box to a large van. The functions ROVs perform can range from simply carrying a camera to observe subsea environments to housing multiple agile arms to perform numerous functions subsea.

remotely 원격으로 **operated** 작용되는 **umbilical** 밀접히 연결된 **manipulator** 조종자 an agent that operates some apparatus or machine **myriad** 무수한 **enable** 가능하게 하다 **indispensable** 필수불가결한 absolutely necessary; vitally necessary **iteration** 반복 **frailty** 약점 **atomic** 원자폭탄의 **sunken** 가라앉은 **preliminary** 예비의 **endeavor** 노력 attempt by employing effort **observe** 관찰하다 **agile** 민첩한, 기민한 moving quickly and lightly

Used primarily for observation and inspection operations, small Electric Vehicles are the tiniest version of ROVs. These robots simply carry a single camera and are deployed in waters no greater than 984 feet (300 meters) deep. High Capability Electric ROVs are larger versions of their Small Electric Vehicle cousins. These ROVs are able to dive to depths up to 20,000 feet (6,096 meters), but they are still only equipped with video cameras and are unable to perform many other tasks because their electricity supply limits them. Primarily used by the science and military industries, High Capability Electric ROVs are unable to perform any heavy tasks typically required by the oil and gas industry.

The next iteration of ROV, the Work Class Vehicle is powered electrically and hydraulically, enabling some more serious work subsea. Although payload and lift capabilities are restricted, the Work Class Vehicle ROV does generally include a seven-function manipulator and a five-function grabber. Most of this class of ROV is deployed in the drilling and construction support sector, as well as pipeline inspection.

The most advanced version of ROV is the Heavy Work Class Vehicle ROV. This kind of ROV is capable of working in waters measuring up to 10,000 feet (3,000 meters) deep. Aptly named, the Heavy Work Class Vehicle boasts horsepower ranging from 100 to 250 and has lift capabilities up to 11,025 pounds. Additionally, these ROVs house multiple manipulators and grabbers capable of performing subsea tie-ins and deepwater installations.

Although the military is testing preliminary versions, Autonomous Underwater Vehicles (AUVs) are the next step in ROV technology and are not yet available in the consumer market. Most likely the next commercial ROV technology will be a hybrid between the AUV and ROV, with less umbilicals needed to connect the subsea robot to its above-water controller. As the oil and gas industry moves to a more subsea environment, ROVs will become an even more important tool in drilling, development and repair offshore. In fact, some ROVs are being built to work on specific subsea developments, with manipulators and tools designed for work on equipment in a precise field. In this case, as the subsea architecture is being designed, the ROV that will handle the subsea work is taken into account, as well, ensuring that all tasks at hand can be managed by the ROV.

deploy 배치하다 to distribute systematically or strategically **task** 직무 **payload** 유효탑재량 **grabber** 잡는거 **aptly** 적절하게 **boast** 자랑한다 speaking of yourself in superlatives, show off **autonomous** 자율적인 not controlled by outside forces **hybrid** 섞인, 잡종의 **precise** 정확한 **take into account** 고려하다 **controller** an electric device used for governing the power that goes to an apparatus to which it is connected

Lecture B

After you read

1. What are the underwater robots that allow the controller to be located above the water?

   (a) Remotely operated robots

   (c) Remotely operated vehicles

   (d) Robot operated vehicles

   (e) Remotely operated vessels

2. Please complete the sentence.

   (a) ROVs are connected to the topside via an __________ link that houses communications cables

   (b) The most advanced version of is capable of working in waters measuring up to __________ feet.

   (c) The Work Class Vehicle ROV does generally include a seven-function manipulator and a five-function grabber. Most of this class of ROV is deployed in the _________ and construction support sector, as well as pipeline inspection.

4. What is not true about the article?

   (a) ROVs are connected to the topside via an umbilical link that houses communications cables, an energy source and information transfer.

   (b) ROVs can only submerge to certain depths, and the skill and danger involved makes deepwater diving unrealistic.

   (c) The first iteration of the robotic diving concept was the Human Occupied Vehicle, or HOV.

   (d) The functions ROVs perform can range from simply carrying a camera to observe subsea environments to housing multiple agile arms to perform numerous functions subsea.

Language Focus

1. Match the words with the definitions. then complete the definition with the given worlds.

HWC ROV	__________robots that allow the controller to be located above the water.
ROV	is powered electrically and hydraulically, enabling some more serious ______________.
WC ROV	house multiple manipulators and grabbers capable of performing subsea __________ and deepwater installations.

underwater tie-ins subsea work

2. Complete the sentences using the following words.

preliminary communications umbilical

1) ROVs are connected to the topside via an _________link that houses ___________cables, an energy source and information transfer.
2) The commercial world quickly began manufacturing _________versions of ROVs for the offshore oil and gas industry.

3. Complete the sentences using the following words.

1) ______, some ROVs are being built to work on specific subsea developments.
   a) In fact b) On fact c) Facting d) To fact
2) While ________ successful, HOVs were limited by human frailties and unable to achieve the water depths needed.
   a) something b) somewhat c) someone

Useful Expressions- Phrasal Verbs Ⅰ

1. Phrasal verb + preposition

get on/off	The passengers can **get on** and **off** the bus easily.
drive off	A man got into his car and **drove off**.
turn around	I **turned around** to see if there was someone behind me. Things **turned around** when he entered the round of 14.
break down	Sorry, I am late. My car **broke down**.
wake up	Don't **wake** the baby **up**.
take off	Sadly, however, Naro lost contact with base 137 seconds after the **take-off.**

run away from	Why are you trying to **run away from** me?
keep up with	Please, walk slowly. I can't **keep up with** you.
look up at	Many people **looked up at** the night sky to see a very special event; a total lunar eclipse!
look forward to	I am really **looking forward to** seeing you again
get along with	I hope I can **get along** well **with** my boss.
run out of	I couldn't call you because my phone **ran out of** batteries.

2. Position of the object as concerned about Phrasal verb

I turned on the light. O   I turned the light on. O

If the object is pronoun, it should come between verb and preposition :
I turned **it** on. (I turned on it X)

Don't { throw away these clothes. / throw these clothes away. } O   Don't throw away them. X

Lecture E	Speak Out!!

1. Talk with your partner. What would you say in these situation? Complete the sentences using the expressions you've studied in this unit.

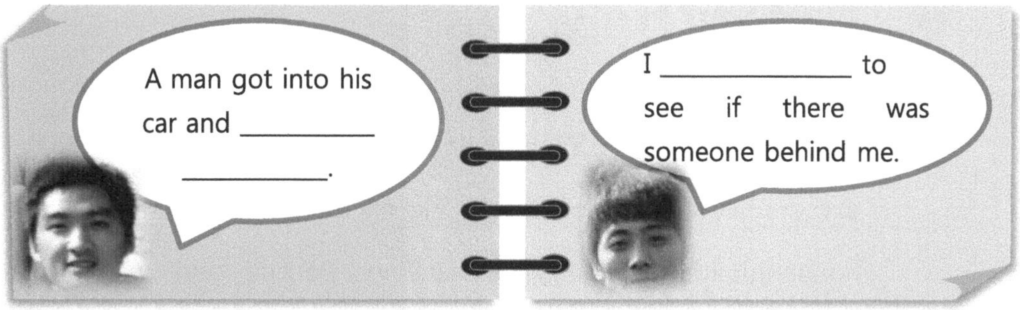

Lecture F	Episode on offshore!!

Work in pair. Ask and answer the question.

	What are ROVs?
	Remotely Operated Vehicles, or ROVs, are underwater robots that allow the controller to be located above the water. ROVs are connected to the topside via an umbilical link that houses communications cables, an energy source and information transfer.
	Why are ROVs indispensable to offshore industries?
	Answer the question!
	What is the most advanced version of ROV ?
	Answer the question!
	*Make your own!*
	*Make your own!*

Offshore Wind Power

Offshore wind power refers to the construction of wind farms in bodies of water to generate electricity from wind. Better wind speeds are available offshore compared to on land, so offshore wind power's contribution in terms of electricity supplied is higher. However, offshore wind farms are relatively expensive.

Siemens, REpower, and Vestas are the leading turbine suppliers for offshore wind power. Dong Energy, Vattenfall and E.on are the leading offshore operators.[1] As of October 2010, 3.16 GW of offshore wind power capacity was operational, mainly in Northern Europe. According to BTM Consult, more than 16 GW of additional capacity will be installed before the end of 2014 and the United Kingdom and Germany will become the two leading markets. Offshore wind power capacity is expected to reach a total of 75 GW worldwide by 2020, with significant contributions from China and the United States.

As of February 2012, the Walney Wind Farm in United Kingdom is the largest offshore wind farm in the world at 367 MW, followed by Thanet Offshore Wind Project (300 MW), also in the UK. The London Array (630 MW) is the largest project under construction. These projects will be dwarfed by subsequent wind farms that are in the pipeline, including Dogger Bank at 9,000 MW, Norfolk Bank (7,200 MW), and Irish Sea (4,200 MW).

Offshore wind power refers to the construction of wind farms in bodies of water to generate electricity from wind. Unlike the term typical usage of the term "offshore" in the marine industry, offshore wind power includes inshore water areas such as lakes, fjords and sheltered coastal areas, utilizing traditional fixed-bottom wind turbine technologies, as well as deep-water areas utilizing floating wind turbines.

Europe is the world leader in offshore wind power, with the first offshore wind farm being installed in Denmark in 1991.[4] In 2008, offshore wind power contributed 0.8 gigawatt (GW) of the total 28 GW of wind power capacity constructed that year. By October 2009, 26 offshore wind farms had been constructed in Europe with an average rated capacity of 76 MW, and as of 2010 the United Kingdom has by far the largest capacity of offshore wind farms with 1.3 GW, more than the rest of the world combined at 1.1 GW[7] The UK is followed by Denmark (854 MW), The Netherlands (249 MW), Belgium (195 MW), Sweden (164 MW), Germany (92 MW), Ireland (25 MW), Finland (26 MW) and Norway with 2.3 MW.

As of October 2010, Danish wind turbine manufacturers Siemens Wind Power and Vestas have installed 91.8% of the world's 3.16 GW offshore wind power capacity, although REpower is now starting to become a major player. Based on current orders, BTM expects 15 GW more between 2010 and 2014.

Pops English

You are my sunshine

The other night dear as I lay sleeping
I dreamed I held you in my arms
When I awoke dear I was mistaken
So I hung my head and cried

You are my sunshine my only sunshine
You make me happy when skies are grey
You'll never know dear how much I love you
Please don't take my sunshine away

So let the sun shine in face it with a grin
Smilers never lose
And frowners never win
So let the sun shine in face it with a grin
Open up your heart
And let the sunshine in

My mommy told me something
That little girls should know
It's all about the devil
And I learned to hate him so
I know he'll be unhappy
Cause I'll never wear a frown
Maybe if we keep on smiling
He'll get tired of hanging round

So let the sun shine in, face it with a grin
Smilers never lose
And frowners never win
So let the sun shine in face it with a grin
Open up your heart
And let the sunshine in

So let the sun shine in, face it with a grin
Smilers never lose
And frowners never win
So let the sun shine in face it with a grin
Open up your heart
And let the sunshine in

The worst is not so long as we can say,
"This is the worst."...

-William Shakespeare

UNIT 16

How
Does the Blowout Control Work?

Blowout Control 시스템은 무엇인가?

Lecture A　How Does the Blowout Control System Work?

If not properly monitored, changes in pressure that can occur while the well is being drilled can cause combustible hydrocarbons to flow unchecked and at high pressures and flow rates. If this flow of hydrocarbons is not stopped in time, the hydrocarbons can ignite into a deadly firestorm called a blowout. Because of the immense cost and danger associated with oil well blowouts, the bulk of the well control industry revolves around the prevention and avoidance of blowouts. Unfortunately, because a blowout only takes a brief moment to occur, it is necessary that there will be methods in place to combat them when the need arises.

In a nutshell, "a blowout is an uncontrolled flow of gas, oil or other formation fluids into the atmosphere or another zone," explained Barry Cooper of Well Control School, an organization, which has offered well control training programs to the oil and gas industry for more than 25 years. "Blowouts are the most tragic and expensive accidents in the upstream petroleum industry," said Cooper. "They endanger life, the environment and future production from the lost well." On an economic level, an oil well gushing thousands or even millions of barrels of oil is costing a company not only in short term production, but also the long-term profitability of the well itself. It is vital to the profitability of the well that the blowout is stopped and the well put back online as quickly as possible. Because of their intensity and the very particular circumstances that set blowouts and oil well fires apart from regular conflagrations, unique and specially trained firefighters must be employed to fix them. Myron M. Kinley, the father of blowout control and oil well firefighting, founded the MM Kinley Company in 1923, which set the trend toward an industry devoted to oil well firefighting. By 1946, the famous oil well firefighter Red Adair had joined the company, and by 1959, he had founded the Red Adair Company.

Today, companies like Boots & Coots, which variously spun off from the Red Adair Company, continue in the same tradition. While the technology has improved significantly over the past 85 years, the basic methods and strategies employed to battle blowouts have remained the same. "The technology may change, but the basics have been around since the `20s, developed by Mr. Kinley and Mr. Adair," said Danny Clayton, Boots & Coots vice president of well control services. The profession is based on tradition, with each generation of oil well firefighters teaching the next generation.

combustible 가연성의 **In a nutshell** 간단히 summed up briefly **formation** 성립, 형성 **immerse** 몰두시키다 **revolve** 회전하다 **prevention** 예방 **avoidance** 회피, 취소 deliberately avoiding; keeping away from or preventing from happening **brief** 간단한 **combat** 전투 **arise** 나타나다 **atmosphere** 대기, 분위기 **tragic** 비극의 **endanger** 위험에 빠뜨리다 **gush** 분출하다 **profitability** 수익성 the quality of affording gain or benefit or profit **vital** 불가결한 **intensity** 강렬 **conflagration** 큰화재, 돌발

Most commonly, when a well is lost to a blowout, the drilling package will have collapsed around the well, making proper assessment of the situation difficult. Firefighters arrive as quickly as possible and use machinery to remove the damaged rig and associated debris so they can assess the situation and choose the best method to fight the blowout. In the early days of fighting oil well fires, the most common technique to smother a blowout was to snuff it with a dynamite blast. Pioneered by Myron Kinley, the intention is to blast fuel and oxygen away from the flame, effectively eliminating the fuel source, similar to snuffing out a candle. Although the first instance of this method dates back to 1913, dynamite blasting continues to be one of the most frequently employed methods.

Another common method employed by oil well firefighters involves drilling a "relief" well or wells into and intersecting the blowing well. This intersection gives the kill fluid a conduit to the surface, enabling what is called a "subsurface kill." A more complicated method for bringing a blown-out well under control involves carefully capping the well with a new blowout preventer, or "BOP." BOPs are essentially large valves on the surface of the well that quickly shut off the well as a last ditch precaution to prevent a blowout from occurring. In this procedure, the detritus of the collapsed rig is carefully removed and a high-pressure abrasive cutter is used to sever the damaged BOP and wellhead for removal. A long boom assembly - at the end of which is a replacement BOP - is maneuvered into position. Large amounts of water are sprayed on the replacement BOP to combat the flames and to keep the replacement BOP from getting too hot. The BOP is quickly lowered onto the well and bolted into place, thus capping the blowout.

Like all forms of firefighting, oil well firefighting requires constant vigilance, attention and teamwork for the safe completion of the job. But sometimes, outside factors complicate the sealing of a well. Not all blowouts necessarily ignite into towering infernos. Sometimes, the hydrocarbons merely blow into the air, which can actually be more dangerous. Often, firefighters will deliberately ignite the blowout as a precaution.

GLOSSARY

collapsed 무너진 an abrupt failure of function or complete physical exhaustion **assessment** 평가 the classification of someone or something with respect to its worth **situation** 위치 **debris** 파편, 잔해 **smother** 은폐하다 질식시키다 **snuff** 심지가 까맣게 탄 부분 **dynamite** 다이너마이트 **pioneer** 선구자 someone who helps to open up a new line of research or technology or art **blast** 폭파 **intersect** 교차하다 **detritus** 파편 **abrasive** 연마재 **removal** 제거 **vigilance** 경계 **infernos** 맹화 **ignite** 불을 붙이다 **precaution** 예방조치

Lecture B **After you read**

1. Who is said to be the father of blowout control and oil well firefighting?
   (a) Barry Cooper
   (c) Myron Kinley
   (d) Danny Clayton
   (e) Barry Clayon

2. What is a blowout ?
   (a) an uncontrolled flow of gas or oil into the atmosphere
   (b) changes in pressure
   (c) combustible hydrocarbons
   (d) High temperature of oil and gas

3. In the early days of fighting oil well fires, what was the most common technique to smother a blowout?
   (a) To snuff it with a dynamite blast.
   (b) To drill a relief well to intersect the blowing well.
   (c) To use dynamite blast to intersect the blowing well.
   (d) To drill the second well to snuff the fire.

4. What is not true about the article?
   (a) When a well is lost to a blowout, the drilling package will have collapsed around the well, making proper assessment of the situation difficult
   (b) After the first instance dates back to 1931, dynamite blasting continues to be one of the most frequently employed methods.
   (c) Sometimes, the hydrocarbons merely blow into the air, which can actually be more dangerous
   (d) oil well firefighting requires constant vigilance, attention and teamwork for the safe completion of the job.

Lecture c	Language Focus

1. Match the words with the definitions, then complete the definition with the given worlds.

Words		Definitions
blowout	●	the quality of affording gain or ________or profit
profitability	●	large ______on the surface of the well that shut off the well as a last ditch _________to prevent a blowout from occurring.
BOP	●	an __________flow of gas, oil or other formation fluids into the atmosphere or another zone

benefit precaution uncontrolled valves

2. Complete the sentences using the following words.

blowout ignite infernos tragic

1) Blowouts are the most _____and expensive accidents in the upstream petroleum industry.
2) The hydrocarbons can ______into a deadly firestorm called a _________.
3) Not all blowouts necessarily ignite into towering ________.

3. Complete the sentences using the following words.

1) ________by Myron Kinley, the intention is to blast fuel and oxygen away from the flame.
   a) Pioneer b) Pioneering c) Pioneered
2) Most ___________ the drilling package will have collapsed around the well.
   a) commonly b) common c) commoned

Lecture D

Useful Expressions- Phrasal Verbs Ⅱ

1. Phrasal verb (verb+ in/out)

In = into a room, building, car, etc.		out= out of a room, building, car, etc.	
get in	I lost my room key, so I can't **get in**.	get out	I don't want to talk to you **Get out**!
check in	As soon as I got to the airport, I **checked in**.	check out	(at a hotel) What time should we **check out**?
move in	A very good-looking boy **moved in** next door!	move out	My best friend **moved out** last month. I feel lonely.

2. The other verbs + in

drop in	= visit s.b for a short time ▪ I just **dropped in** your house if you are ok.
hand in turn in	= give s.t written to a teacher, boss, etc. ▪ Your homework is due this week. Don't forget to **hand** it **in** by this Friday.
fit in	= feel you belong well by a group ▪ He doesn't seem to **fit in** with the rest of the kids at school.

3. The other verbs + out

fill out	a form, application, etc. = write the required information onto a form. ▪ I have to **fill out** this visa application by the end of this month.
drop out	of school/ college/ a course, etc=quit before completion ▪ Mike went to university but he **dropped out** after a year.
figure out	= understand the meaning ▪ Did you find the solution? No, I couldn't **figure** it **out**.
find out	= get to know The police never **found out** who committed the murder.

Lecture E **Speak Out!!**

A. Talk with your partner. What would you say in these situation? Complete the sentences using the expressions you've studied in this unit.

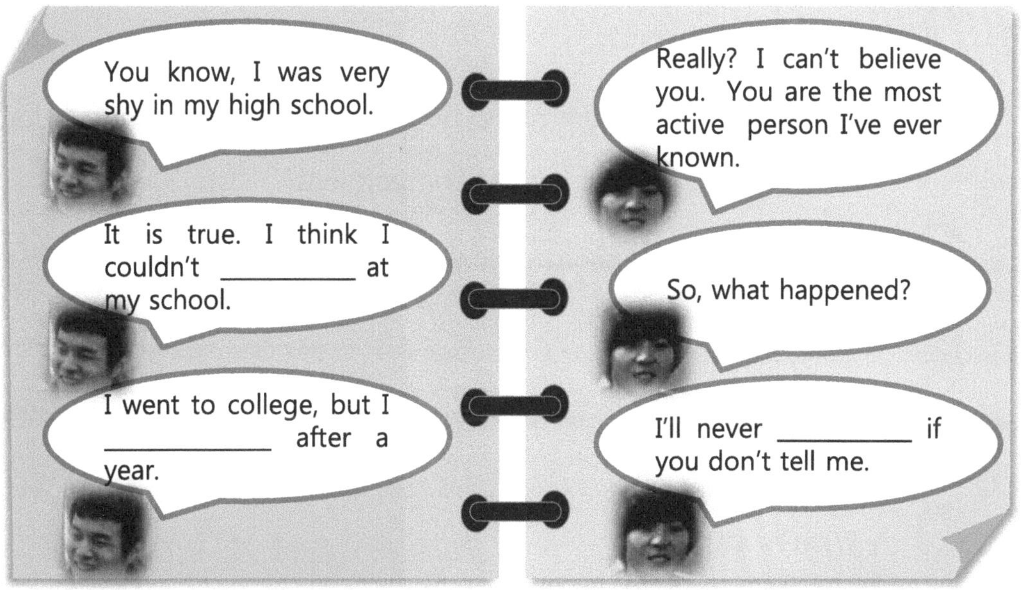

Lecture F

Episode on offshore!!

Work in pair. Ask and answer the question.

	What is a blowout?
	A blowout is an uncontrolled flow of gas, oil or other formation fluids into the atmosphere or another zone
	What is the consequence caused by the blowout?
	Answer the question!
	What is a BOP?
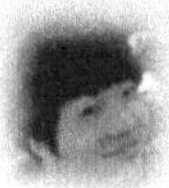	Answer the question!
	*Make your own!*
	*Make your own!*

OUTRO

Deepwater Horizon Oil Spill

On April 20, 2010, an explosion and fire occurred on the Deepwater Horizon drilling rig in the Gulf of Mexico. This resulted in 11 worker fatalities, a massive oil release, and a national response effort in the Gulf of Mexico region by the federal and state governments as well as BP.

Based on estimates from the Flow Rate Technical Group, which is led by the U.S. Geological Survey, the 2010 Gulf spill is the largest oil spill in U.S. waters. The oil spill has damaged natural resources and has had regional economic impacts. In addition, questions have been raised as to whether offshore regulation of oil exploration has kept pace with the increasingly complex technologies needed to explore and develop deeper waters.

Crude oil has been washing into marshes and estuaries and onto beaches in Louisiana, Mississippi, and Alabama. Oil has killed wildlife, and efforts are underway to save oil-coated birds. The most immediate economic impact of the oil spill has been on the Gulf fishing industry.

The commercial and recreational fishing have faced extensive prohibitions within the federal waters of the Gulf exclusive economic zone. The fishing industry, including seafood processing and related wholesale and retail businesses, supports over 200,000 jobs with related economic activity of $5.5 billion. Other immediate economic impacts include a decline in tourism. On the other hand, jobs related to cleanup activities could mitigate some of the losses in the fishing and tourism industry.

The Bureau of Ocean Energy Management, Regulation, and Enforcement (BOEMRE), formerly known as the Minerals Management Service (MMS), and the U.S. Coast Guard are the primary regulators of drilling activity. The Coast Guard generally overseas the safety of systems at the platform level of a mobile offshore drilling unit. The Environmental Protection Agency (EPA) has multiple responsibilities, with a representative serving as the vice-chair of the National Response Team and Regional Response Teams. The Federal Emergency Management Administration has responsibilities with respect to the economic impacts of the spill; its role so far has been primarily that of an observer, but that may change once the scope of impacts can be better understood.

BOEMRE/MMS is also the lead regulatory authority for offshore oil and gas leasing, including collection of royalty payments. Its regulations generally require that a company with leasing obligations demonstrate that proposed oil and gas activity conforms to federal laws and regulations, is safe, prevents waste, does not unreasonably interfere with other uses of the outer continental shelf, and does not cause impermissible harm or damage to the human, marine, or coastal environments. Further attention to the internal organization of BOEMRE/MMS is an ongoing legislative and regulatory focus.

Pops English

Let It Be

When I find myself in times of trouble
Mother Mary comes to me
Speaking words of wisdom
Let it be.
And in my hours of darkness
She is standing bright in front of me
Speaking words of wisdom
Let it be.
Let it be.
Let it be.
Let it be.
Let it be.
Whisper words of wisdom
Let it be.
When the broken hearted people
Living in the world agree
There will be an answer.
Let it be.
Though they may be parted
There is still a chance that they will see
There will be an answer
Let it be.
Let it be.
Let it be.
Let it be.
Let it be.
There will be an answer.
Let it be.
When the night is cloudy
There is still a light that shines on me.
Shine until tomorrow
Let it be.
I wake up to the sound of music
Mother Mary comes to me
Speaking words of wisdom.
Let it be.
Let it be.
Let it be.
Let it be.
Let it be.
Whisper words of wisdom
Let it be.

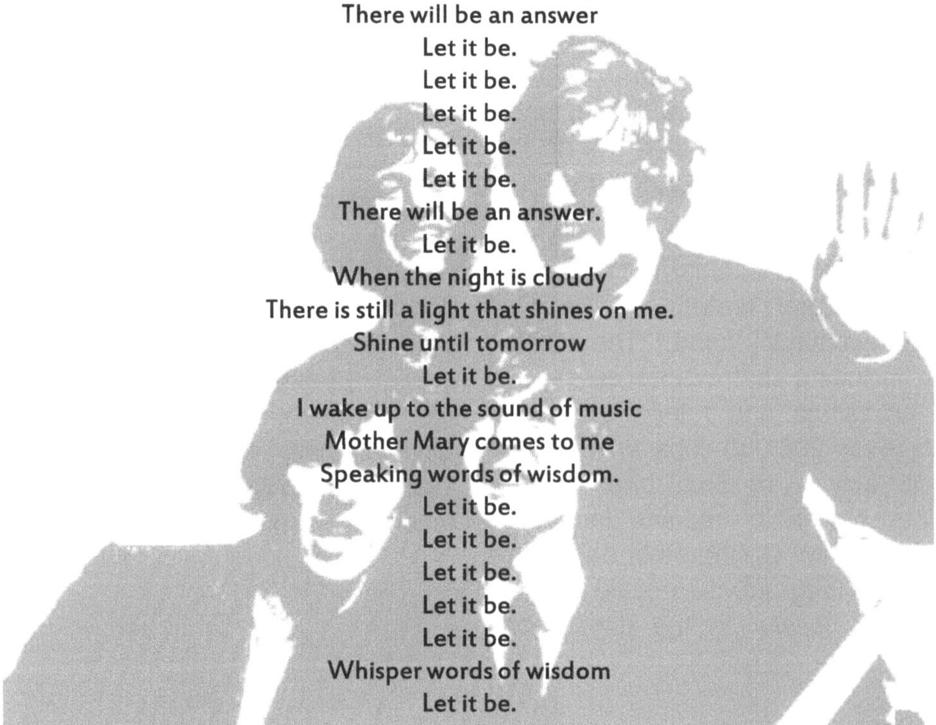

It's not that I'm so smart , it's just
that I stay with problems longer.

- Albert Einstein

Reference

최경식, 2010, 해양플랜트개론, 문운당

김태희, 손호재, 박창수, 2007 해양플랜트 공학, 선학출판사

Sources from internet

Unit 1	**Reading** http://en.wikipedia.org/wiki/Drill **Outro** http://www.petroleumhistory.org/OilHistory/pages/Whale/whale.html
Unit 2	**Reading** http://www.rigzone.com/training/insight.asp?insight_id=299&c_id=12 **Outro** http://en.wikipedia.org/
Unit 3	**Reading** http://www.rigzone.com/training/insight.asp?insight_id=299&c_id=12 **Outro** http://en.wikipedia.org
Unit 4	**Reading** http://www.rigzone.com/training/insight.asp?insight_id=339&c_id=24 **Photoes** http://www.2b1stconsulting.com/tlp/
Unit 5	**Reading** www.globalsecurity.org **Outro** http://en.wikipedia.org
Unit 6	**Reading** www.globalsecurity.org/ **Outro** http://en.wikipedia.org
Unit 7	**Reading** http://www.rigzone.com/training/insight.asp?insight_id=306&c_id=24
Unit 8	**Reading** http://www.marin.nl/web/Ships-Structures/Merchant-vessels-Work-boats/Shuttle-Tankers.htm
Unit 9	**Reading** http://en.wikipedia.org **Outro** http://en.wikipedia.org
Unit 10	**Reading** http://www.rigzone.com/training/insight.asp?insight_id=311&c_id=19 **Outro** http://en.wikipedia.org
Unit 11	**Reading** http://www.rigzone.com/training/insight.asp?insight_id=358&c_id=17 **Outro** http://en.wikipedia.org
Unit 12	**Reading** http://gcaptain.com/history/ , http://www.marineinsight.com/marine/types-of-ships-marine/what-is-a-dynamic-positioning-ship/
Unit 13	**Reading** http://www.2b1stconsulting.com/turret/ **Outro** http://worldoceanreview.com/en/energy/methane-hydrates/
Unit 14	**Reading** http://www.ivt.ntnu.no/ept/fag/tep4215/innhold/LNG%20Conferences/2005/SDS_TIF/050155.pdf **Outro** http://blog.liverpoolmuseums.org.uk/default,date,2012-01-05.aspx
Unit 15	**Reading** http://www.offshore-technology.com/features/feature1659/
Unit 16	**Reading** http://www.rigzone.com/training/insight.asp?insight_id=300&c_id=1

Offshore English
쉽게배우는 해양실무영어

2012년 9월 5일 초판 인쇄

2014년 1월 10일 2 쇄 발행

저 자 이정은

발 행 인 송기수

발 행 처 도서출판 GS인터비전

편 집 처 도서출판 GS인터비전

인 쇄 처 대일문예사

등록번호 제 25100-2016-000050 호

I S B N 978-89-97231-53-9 (93740)

주 소 서울 은평구 증산로15길 69, 2층

전 화 02-976-7898, 02-3272-7898. 0707-720-7898

팩 스 02-6468-7898

홈페이지 gsintervision.co.kr

E-Mail gsinter7@gmail.com

정 가 15 ,000원